# VANISHING LIFE ON EARTH



BIMALENDU B. NATH

BlueRose ONE.com
Stories Matter
New Delhi • London

**BLUEROSE PUBLISHERS**
India | U.K.



BlueRose ONE.com
Stories Matter
NewDelhi • London

For permissions requests or inquiries regarding this publication, please contact:

BLUEROSE PUBLISHERS
www.BlueRoseONE.com
info@bluerosepublishers.com
+91 8882 898 898
+4407342408967

ISBN: 978-93-6452-093-5

Cover design: Yash Singhal
Typesetting: Namrata Saini

First Edition: January 2025

***Dedicated***

*to*

all my colleagues, friends and students

who engaged me in exploring the evolution of life

*"To understand the history of life, we need to understand extinction."* --- Selina Cole

****

*"Extinction is the rule; survival is the exception"* --- Carl Sagan

****

*"The fate of the vast majority of species on this planet has been extinction, eventually"* --- Alice Roberts

****

# Contents

# *Acknowledgements*

At the outset, I acknowledge many of my students who encouraged me to put my evolutionary ideas and lectures down on paper and compile them into a book on the history of life on Earth in layman's terms. Nevertheless, every single-authored book finally gets published, with supporting roles played by friends, colleagues, and well-wishers. I am privileged to receive help from many eminent personalities, and it is a pleasure to acknowledge all of them.

I am grateful to Prof. Jack W. Szostak, Nobel prize laureate (*Professor at the University of Chicago and former Professor in Genetics at Harvard Medical School*), for sharing and permitting me to reproduce images from his work. His encouraging messages wishing success for the book will remain an asset to me.

I sincerely thank Prof. J. William Schopf (*University of California, Los Angeles*) for providing a representative image of a microfossil he discovered in a rock sample from the West Australian deposits.

My special thanks go to Ms Coco McGrath and *Bush Heritage, Australia*, for allowing me to use an image of Stromatolites, the ancient 3.45 billion years old fossilized microbial mats.

I thank Dr. Valeria Lencioni for her advice while desperately searching for an image of a Cenozoic fossil. I am also thankful to her for providing all assistance during my visit to the museum at *MUSE-Museo delle Scienze, Trento, Italy*, to explore prehistoric collections of fossils.

My special thanks to my friend Dr Andrey Przhiboro (*Zoological Institute, St. Petersburg, Russia*) for enriching my knowledge of fossils of Chironomid midges whenever we met in person and during hiking trips. I especially thank him for providing an image of the fossil imprint of the Chironomid midge.

I thank Prof. Wojciech Gilka (*University of Gdańsk, Poland)* for providing a fossil image of a 'mime' midge. I also thank Dr. Viktor Baranov (*Doñana Biological Station—CSIC, Seville, Spain*) for sharing a photo of an adult insect fossil. I am grateful to Dr. Martin Spies (*SNSB—Zoologische Staatssammlung München, Germany*) for responding to my SOS emails whenever I needed help while writing this book.

I would like to acknowledge other eminent academicians and scientists whose teachings and writings have significantly influenced my thinking while writing this book. First, I must mention Prof. Samir Bhattacharya and Prof. Shelley Bhattacharya from Visva Bharati Central University in Santiniketan, India. As mentors, they profoundly shaped my intellectual development and guided me in interpreting biological phenomena through an evolutionary perspective. During my formative years, a notable public lecture by Prof. Samir Bhattacharya titled "Living and Life" instilled the philosophy of 'adapt, evolve or perish' in my young mind.

Next, I express my gratitude to Prof. Subhas Chandra Lakhotia at Banaras Hindu University, Varanasi, India, who was my PhD advisor. He provided me with the opportunity to investigate the expression of 'heat-shock' genes in an evolutionarily ancient midge insect that evolved during the Jurassic era (approximately 200 million years ago). This experience marked my initial encounter with what is often referred to as a 'living fossil'.

My heartfelt thanks to Prof. Raghavendra Gadagkar, Indian Institute of Science, Bengaluru, India, for many engaging discussions whenever our paths crossed in person. His books, articles, and popular lectures have continually inspired me, and readers may find echoes of these influences in this book.

At the time of writing this acknowledgement, I remember my friends and colleagues at the Association of Teachers in Biological Sciences (ATBS), especially Dr. Rekha Vartak, Dr Anumapa Ronad, Prof. Madan M Chaturvedi, Prof. Anindya Sinha, Prof. P.G.Kale, Dr Sasikumar Menon for stimulating discussions during several Resource Generation Camps at *Homi Bhabha Centre for Science Education (TIFR), Mumbai, India.* Such brainstorming sessions have been held yearly and have provided impetus for selecting sub-topics and narratives described in this book. I thank Dr Rekha Vartak, Dr Anupama Ronad and Mr Vikrant Ghanekar for inviting me to deliver lectures to nationally selected young biology students and mentoring the best ones during the Biology Olympiad training camps. These learning experiences helped me decide what must be included in the book.

I would like to convey a very special and heartfelt thanks to Prof. Suresh Gosavi, Vice Chancellor, Savitribai Phule Pune University, Pune, India; Mr. Yasir Nainar, President; and Mr. George Thomas, Vice President of MIE-SPPU Institute of Higher Education, Doha; Qatar. Their commitment to cultivating an exceptional academic environment has been instrumental in my ability to educate and shape young minds on a few value education courses focusing on environmental sustainability, climate change and conservation, all of which have played a vital role in shaping this book.

I would also like to express my deep appreciation for all my colleagues within the Department of Zoology at SPPU in

Pune, India. Special thanks go to Prof. Narahari Gramapurohit and all past department heads for believing in my pedagogy whenever I designed and updated courses centered around animal evolution and diversity.

I owe a lot to my Astrophysicist brother, Prof Biman Nath, a cosmologist who taught me the secrets of the Universe and the evolution of galaxies and stars in layman's terms. For that matter, he taught our entire extended family stargazing! Over the years, our thought-provoking conversations about life's evolution, fossils, and mass extinctions have profoundly shaped my perspectives on our existence, the planet, and the vast universe beyond.

I acknowledge unwavering support from my loved ones: my wife Anurita and son Anuraj. I am particularly thankful to Anurita for her meticulous copyediting of the manuscript and to Anuraj for his insightful suggestions regarding illustrations and aesthetics. Their comments were crucial for me to improve the book's quality at various writing stages.

Finally, I would like to thank all my past and present PhD students, particularly Dr Ananda Babrekar, Dr Kailas Datkhile, Dr Rahul Gaikwad, Dr Leena Thorat, Dr Pallavi Gaikwad, Dr Anupam Ronad, Dr Kanchan Phatak, Dr Pratibha Bomble, Ms. Minal Ronge, Ms. Chaitali Shinde, and Ms. Snehal Ladkat, who happily agreed to add questions in their research that needed an evolutionary perspective, and it paid off for this book!. I especially hope that the general readers will find the book worth reading.

---- *Bimalendu B. Nath*

# *Preface*

I remember three special occasions that drew my attention to the past life on Earth. During my childhood days, I used to go hiking with my father. My father started his career as a civil engineer sometime during the 1950s, and he was responsible for constructing hill roads and tunnels in the interior of Meghalaya, the North-Eastern part of India. He was my first teacher of Natural Sciences, and he taught me the language of forests and wildlife. On several occasions, he found many fossils near the construction sites in the west Garo hills, India. The fossil stories shared by my father used to be like fairy tale stories, which any child would be fascinated with. That was the beginning of my romance with the past footprints of life on Earth. However, I was not happy with the limited and isolated information on fossils, the sudden disappearance of animals, and the geo-climatic features of the primitive Earth in our textbooks available in the 1970s and 1980s. It was difficult to connect the dots because I was not studying Earth Sciences as my core subject.

My hunt for an 'all-in-one' book on the history of life on the library shelves began on the day I graduated. Needless to say, I started searching for an author since then. Much later, when I was pursuing my PhD on gene expression under environmental stress, my mentor, Prof. Subhash Chandra Lakhotia, called me one day to extend my doctoral work on an ancient midge insect, *Chironomus*, which evolved around the Jurassic era and whose lineages have thrived till day. He helped me to align my thought process with the evolutionary perspective. This sudden motivation helped me reopen the history book of life on Earth. While reading about the ancient midges, I started looking at the Earth's deep past. Once again, I was disappointed to find a void in the popular science books referred to non-specialists where emphasis is mostly given to how and when a species originated rather than how (and why) the species vanishes forever. No wonder the famous fossil biologist David Raup once wrote, *"The extinction of species is not normally considered an important element of Neo-Darwinian theory, in contrast to the opposite phenomenon, speciation"*. Ironically, I learned more about species' origin than species' disappearance. Nevertheless, I used to refer to books on Geology and Earth sciences to link one species' extinction phenomena to another in the backdrop of Earth's geophysical and climatic conditions. I was desperate to search for someone who could write a book merging astrobiology, geology, climatology, chemistry, zoology, botany and microbiology to narrate the story of life on Earth. My search continued until I entered my professional career and started teaching evolutionary biology, biodiversity, genetics and allied subjects in the Department of Zoology, University of Pune (now Savitribai Phule Pune University). Finally, I decided to write a concise book all by myself. Although teachers usually motivate and inspire their students, in my case, my students provoked me with their queries to relook into the history of life and inspired me to write a book in layman's terms. This book is a culmination of my interactions with students, colleagues, friends, and scholars over the years.

Years back, another impetus for writing this book came one evening when I went to a theatre with my family to watch Steven Spielberg's blockbuster movie *Jurassic Park*. Incidentally, I met a few friends who came to watch the movie with their school-going children, and I was introduced to these young kids as the most knowledgeable person about fossils and extinct animals! Before I could digest such a hyped profile about myself in superlative words, the children started firing me with questions like, "*Our teacher told us the story of the movie is fiction, and no dinosaurs can be found now except their skeletons. Is that true?*" "*Do you know who killed all the dinosaurs*"? "*Where were humans when Dinosaurs died?*" and so on. The final blow came when my friend's wife desperately asked me, "*Do you know any Zoo where I can take my son to show a live dinosaur*"? At that moment, I invited them for a post-movie coffee treat and began my classroom-type narration on dinosaurs and the story of their extinction. The discussion ended with a desperate question from my friend's daughter: "If not on Earth, dinosaurs must be found on any distant planet!". Stevan Spielberg successfully added the term 'Jurassic' into the vocabulary of the common man who may not even know the geological calendar of Earth.

Nevertheless, during my classroom interactions with the students, I discovered many 'myths' associated with the animals that permanently vanished from the Earth. We often fail to appreciate the fact that the evolutionary history of life is more of a stochastic phenomenon – influenced by chance events in nature rather than a pre-determined journey. I tried to emphasise this while dealing with the five big mass extinctions that reshaped life and Earth over hundreds of thousands of years of its existence. I attempted to open a window to the past when our Earth and life forms were different, and none exist anymore. Lastly, I would like to tell my readers that the book mainly focuses on animals that lived in the past and have permanently vanished from the Earth and also on those animals struggling to thrive under the human-driven era of climate change.

--- *Bimalendu B. Nath*

# CHAPTER 1

# *Origin of Life*

## The missing pieces

Since the day we learnt to distinguish living and non-living entities around ourselves, our curiosity arose to know how living forms exist. When we realise we are one of those living entities, we wonder how life came onto earth, in what forms life existed in the beginning and so on. However, there was no life on Earth for millions of years, and it was a 'naked planet'. Almost 4.54 billion years ago, Earth and other planets, satellites, and asteroids of the solar system came into existence only as a huge cloud of dust and gas. In the earliest formative phase, dust and gas coalesced due to gravitational pull, and that's how the journey of our planet Earth began, similar to other counterparts.

So far, scientists have not found any evidence of 'life' elsewhere in the Universe, and therefore, the presence and origin of life on Earth have remained one of the intriguing riddles in science. Many of us would be satisfied with an answer that life arose "by accident". However, what an incredible 'accident' to imagine, one that occurred on primaeval Earth and how life evolved after that! Similar to the origin of the first signature of life, many stochastic events took place one after the other that gave rise to the first 'cell'. Eventually, modern life forms came into existence through hundreds of millions of years of evolution. This chapter is the story of these dramatic events on Earth. Myriads of organic and inorganic molecules underwent physicochemical processes leading to the evolution of the first cell – the building block of life.

## The early Earth before life began

If you can go back in time hypothetically, when Earth was very young and return to the present-day Earth, you will encounter a contrasting atmosphere. At present, 99% of Earth's atmosphere is composed of nitrogen ($N_2$) and oxygen ($O_2$), the rest being carbon-di-oxide ($CO_2$), inert gases, water ($H_2O$) vapour etc. On the contrary, scientists revealed that Earth's early atmosphere was predominantly (almost 97%) $CO_2$, the rest being $N_2$ and water vapour. All the oxygen atoms were initially 'locked' in $CO_2$ and $H_2O$ vapour till the Earth gradually oxygenated. It is, therefore, deduced that the primitive version of life forms first arose when there was no free oxygen, and it did not require oxygen to sustain and grow. Other features of the primitive atmosphere do not exist anymore. The primitive surface of Earth was frequently bombarded by high doses of harmful ultraviolet radiation, which does not occur in present-day Earth due to ozone layer shielding. Most of the Earth was subjected to frequent collisions with asteroids and extreme volcanic activity. The gases released from the Earth's interior changed the primordial atmosphere, especially carbon dioxide, water vapour, sulphur dioxide, and nitrogen from volcanic degassing. Subsequently, with frequent volcanic activities, the atmosphere was added with hydrogen, carbon monoxide, methane, ammonia, and hydrogen sulphide. Over time, the Earth's surface cooled down and solidified, forming a solid crust. Gradually, the water vapour condensed, and oceans were formed through centuries of rain filling up all water bodies. Although we can conceptualise many past episodes with plausible answers, life's origin on Earth has remained debated. The most widely accepted assumption is that life originated on Earth through a physicochemical process. The refinement of scientific knowledge has given rise to many theories. Let us look at a few of these hypotheses on the origin of life on Earth.

## Land or sea: Where did life come from?

One of the hypotheses suggested that frequent cloud-to-ground lightning strikes might be crucial in forming key organic molecules from nitrogen and carbon for life to kick-start on Earth. Formation of key organic molecules was otherwise impossible, and presumably, a primordial form of life arose from these raw materials through a series of progressive chemical reactions. These extraordinary reactive processes were possible in the preponderance of ultra-violet components of solar radiation coupled with electric discharges in the form of lightning. This hypothesis received experimental support from a simulated environment, which will be discussed later in this chapter. Once synthesized in the atmosphere, the pool of organic molecules rained down into the ocean (Figure 1.1)

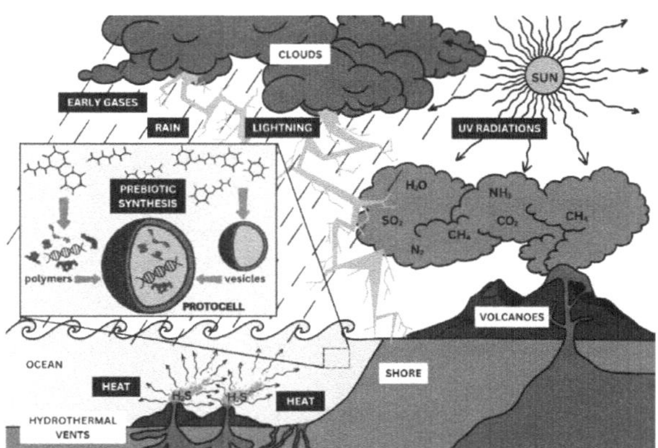


Figure 1.1: Schematic representation of early earth environment and proposed scenarios for the formation of biomolecules: Hypothetical depiction of various theories to explain the origin of life. [*Illustration credit: Uzma Shaikh, TIFR Mumbai, India*]

## Did life begin in the deep sea?

Another hypothesis proposes that biologically relevant organic molecules originated in shallow pools of water, where conditions were supposedly congenial and stable for life to begin. This hypothesis contrasts with another theory that supports the beginning of life in the deep ocean. Nevertheless, the popular theory of oceanic sites as the origin of primordial life gathered much evidence, including fossils recovered from hydrothermal vents. What are these hydrothermal vents? Modern-day Earth contains faults or cracks beneath the crust, and these massive irregularly shaped solid slabs of rock are called a tectonic or continental plate, which periodically moves either towards or away from each other due to pressure built up from the Earth's interior. Generally, earthquakes originate from the region of fault lines lying between the tectonic plates. Hydrothermal vents can be found in such locations, and these vents eject minerals, dissolved gases and the most crucial elements of life, such as carbon and hydrogen. The energy from the heat escaping from the Earth's mantle is believed to trigger the formation of the first organic molecules comprising carbon and hydrogen, essential for the primordial life forms.

Figure 1.2: Example of a hydrothermal vent as it exists today under deep sea

## Was life seeded from outer space?

An altogether different theory, called panspermia, gained support from some scientists who postulated that cosmic bodies like asteroids, meteorites or even comets hit Earth quite often and brought rocks from outer space, thereby seeding some form of life. However, this theory neither answers how life originated on any other planet or in other celestial bodies nor can it satisfy the doubt of how the alien form of primitive life could thrive in space before landing on Earth! However, scientists reported the presence of complex organic compounds in outer space. Different amino acids, building blocks of proteins, were found in meteorites that fell on Earth. Proponents of the extra-terrestrial origin of life gathered evidence supporting the idea that raw materials of primordial life came from outer space.

## Chemical clue to the origin of life

No one can deny that the story of life on early Earth has several unanswered questions. However, the real question was how the origin of biomolecules, specifically the long biomolecule chains known as biopolymers, came into existence. At this point, one needs to ponder how laws of chemistry and physics were operating on the so-called "broth of molecules" in the hot oceanic water under extraordinary conditions on the early Earth. Many scientists agree on the occurrence of serendipitous 'chemical accidents', which facilitated chemical reactions of the right combination of molecules forming biomacromolecules. It is widely accepted that these molecules came together purely by chance. This explanation was first given by the Russian biochemist Alexander Ivanovich Oparin, and often, it is referred to as the Chemosynthetic theory of the origin of life. A similar postulation was also given by J.B.S. Haldane, suggesting the origin of biomolecules by a combination of various chemical substances through a series of chemical reactions under extraordinary conditions.

Oparin's chemosynthetic theory of the origin of life usually merges with Haldane's and is better known as the Oparin-Haldane hypothesis. Let us look at these events in detail through the viewpoints of the Oparin-Haldane hypothesis and Miller-Urey's experiments. According to Oparin, life originated from 'coacervates', which are microscopic clusters of lipids (fat) molecules bound by electrostatic forces, and Oparin proposed that these 'protocells' might be the precursors of complex cells as they exist now. Oparin also experimented with enzymatic reactions – a fundamental step for complex metabolic reactions to evolve later. The Interior of coacervates was populated by simple molecular precursors (monomers) of complex macromolecules (polymers), which could undergo self-assembly and carry out catabolic (breaking down) and anabolic (synthesis) type of metabolic reactions in the presence of enzymes. Oparin envisaged how the first protocell would have thrived in hot oceanic water, metaphorically referred to as "primordial soup".

Although Oparin and Haldane's viewpoints were questioned for a few specifics, it is now generally accepted that there has been a gradual transition from simple inorganic to organic and simple organic to complex organic macromolecules. Present-day categories of biomolecules are polymers, which are long chains of organic molecules. To begin with, carbon combined with itself to make long chains (a property called 'catenation'). These long chains of carbon combined with hydrogen formed hydrocarbons, giving rise to key organic molecules like aldehydes, ketones, acids, etc. Attainments of life's primary requirements for carbon-based molecules were major advancements towards continuing primordial life to modern life forms.

## Simulation of primitive Earth and the birth of Exobiology

In modern life forms, there are four basic biological macromolecules: carbohydrates, lipids, proteins and nucleic acids. The major steps of the chemical evolution of life were marked by the formation of precursor biomolecules, which went through polymerisation to form complex macromolecules as we know them today. Carbohydrates or polysaccharides were formed by polymerising sugar monomers (monosaccharides). Similarly, lipids or fats were formed from fatty acid monomers. Roles of proteins are multifarious, including enzymes, which are proteins. Proteins are polymers made of amino acids as monomers.

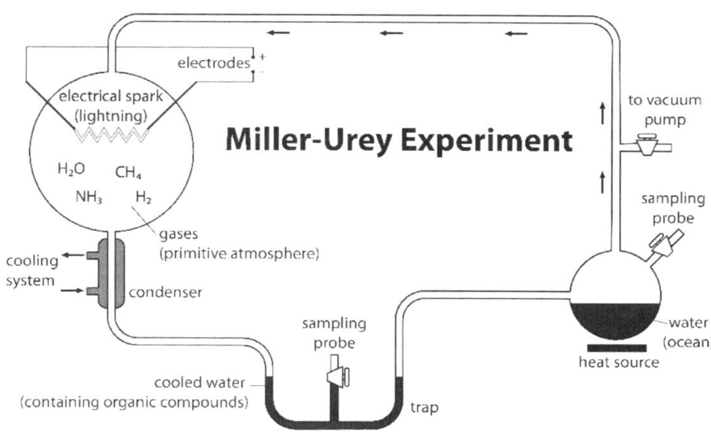


Figure 1.3 : Schematic representation of Miller-Urey simulation experiment.
[Illustration credit: Dr Vikas Kashid, MIE-SPPU IoHE Doha Qatar]

DNA and RNA are nucleic acids, known as life's genetic material, and comprise nucleotides as monomers. Thus, a chain of molecules of similar chemical properties linked together, giving rise to different categories of biological polymers, and life's basic chemical ecosystem was established.

However, Miller and Urey tested Oparin-Haldane's proposed concept of the origin of biomolecules through simple natural chemical processes. Miller-Urey's simulation experiment provided the first evidence that organic molecules could be formed out of inorganic molecules under the extraordinary conditions of primitive Earth outlined in the Oparin-Haldane hypothesis.

On May 15, 1953, the journal SCIENCE published an experiment which eventually gave birth to a branch of biological sciences: Exobiology or Astrobiology (the study of life in the Universe). Stanley Miller (under his University of Chicago mentor & Nobel laureate Harold Urey) carried out the classic experiment. Miller-Urey's experiment aimed to test the central part of Oparin-Haldane's hypothesis by simulating the environmental conditions that prevailed on the early Earth. The objective was to determine whether organic molecule formation from inorganic compounds was possible. The experimental setup comprised a gaseous chamber with methane, ammonia, and hydrogen. They recreated Earth's hostile atmosphere using electric sparks. They simulated lightning as an energy source inside the gaseous chamber (Fig. 1.3). A separate chamber was kept with water boiling inside, which mimicked a primitive ocean. Water vapours from this oceanic chamber moved towards the atmosphere, filled with gases representing a primitive atmosphere. The experimental set-up further included a condenser. The condensed liquid sample was collected and chemically examined within a week. The analysis revealed the presence of different types of organic molecules, including amino acids, which are the building blocks of proteins in all living organisms. In subsequent years, Oparin-Haldane's hypothesis has been continually tested, and Urey-Miller's experiment was repeated by others several times using advanced tools and techniques. The results were reproducible, and many additional amino acids and organic compounds were detected.

## Candidates to be life's first set of molecules

There are two essential criteria which need to be addressed to study the early form of life: namely, the acquisition of one group of molecules to drive chemical reactions as catalysts (modern-day enzymes) and another group of molecules with self-replicating properties (modern-day deoxyribonucleic acid, i.e. DNA1 and ribonucleic acid, i.e. RNA). The first group of molecules involves 'metabolism', one of the living entity's most crucial defining features. All chemical reactions occurring in any living being are collectively called metabolism. It comprises 'anabolism' (synthesis of formation of complex macromolecule from simple one) and 'catabolism' (breaking down complex macromolecules into simple ones). Catalysts are required to facilitate metabolic reactions, and enzymes do this job. The question is, had there been enzyme-like molecules at the beginning of life on Earth? Before going into details of molecules functioning as enzymes, which are basically proteins, the presence of another category of molecules is essential for life to perpetuate – the self-replicating molecules. Origin of life researchers have a dilemma about which came first: proteins (serving as enzymes) or nucleic acids like DNA-RNA (serving as genetic materials). This is like an old-time riddle: which came first—the chicken or the egg? Similarly, researchers debated whether the 'protein-first' or 'DNA-first' theory should be accepted.

## 'Chicken and egg' riddle: The RNA world

According to the RNA world hypothesis, RNA was the primary living substance available during the early phase of life, largely because RNA can function as both genes and enzymes. The discovery of ribozymes – a class of RNA capable of catalysing chemical reactions as enzymes and function as genetic materials. Many ancient viruses have RNA as their sole genetic material. These logics have strengthened the RNA-world hypothesis and gained prominence amongst other alternative theories. Researchers started pondering whether RNA might have come before DNA and protein because RNA, as the primordial molecule, satisfies the basic qualification of life, driving hundreds and thousands of chemical reactions and engaging in self-replication to propagate life's primordial forms.

Oparin-Haldne's hypothesis could not solve the proverbial 'chicken-and-egg' riddle in the chemical evolution of life. A functional living cell essentially needs a self-replicating molecule like DNA, which stores genetic information, and thereby, one cell reproduces to give rise to two identical cells, and two give rise to four, four to eight. A living entity can perpetuate like this. However, without protein enzymes, DNA cannot be replicated. On the other hand, proteins are encoded in DNA and cannot be formed without DNA being around. So, which comes first, Protein or DNA or vice-versa? The riddle was solved by the 'RNA world's hypothesis, which proposed that RNA evolved before DNA and protein, and now let us talk about it in little detail. The discovery of 'ribozyme' in 1982 convinced the origin-of-life researchers that RNA originated before DNA and proteins. Ribozymes can catalyse a specific chemical reaction similar to what is known for protein enzymes. Hence, ribozymes are called ribonucleic acid (RNA) enzymes. Secondly, the discovery of ribozyme helped us to recognize that RNA might have evolved before DNA. Another phenomenal discovery of ribozymes as a self-replicator in 2002 finally ended the 'chicken and egg' debate of the origin of life as to which came first: protein or DNA?

## Connecting dots --- linking missing pieces

Now, to conclude the whole story of the origin of life on Earth, whatever had been the mode of beginning--- atmospheric, in deep-sea hydrothermal vent or seeded from outer space, life continued to thrive on Earth because of organic entities containing self-replicating molecules. The coacervate-like lipid-layered vesicles, having RNA and other metabolically active polymers and enzymes to carry out chemical reactions, became the precursor to 'protocells' or the first living cells. The protocells provided an isolated chemical environment because these are bound by lipid layers which are hydrophobic (repellent to water). These vesicles grew in ocean water by gradual accumulation (known as accretion) of nearby lipid vesicles or micelles. Proponents of the RNA world hypothesis believe that DNA evolved from RNA much later in the early history of light. Subsequently, the perpetuation of life was driven by evolutionary processes and led to the origin of the first cells equipped with hereditary, metabolic and cellular reproductive systems.

Over the last few decades, many theories for the origin of life have emerged. Oparin-Haldane's concept of early Earth was questioned, and the Earth's early atmosphere is no longer believed to prevail as chemically reducing. However, Oparin-Haldane's proposed reducing environment can be found in deep-sea hydrothermal vents, and ambient heat surrounding the vents could have served as the energy source. Although part of Oparin-Haldane's view was contested, the concept of life's origin through chemical evolution has remained widely accepted. Firstly, Miller-Urey's experiment convincingly demonstrated how complex organic molecules can be produced from inorganic molecules under extraordinary conditions on primitive Earth. Secondly, it is generally accepted that 'coacervate' like droplets served as microscopic compartments that could separate internal and external environments, ensuring that the organic material within the coacervate was not lost in the ocean. Thirdly, coacervates possessing DNA/RNA emerged as the first 'living cells'.

# CHAPTER 2

# *Diversification of Life*

In life that exists on Earth today, all cells are composed of organic molecules made of CHNOPS (Carbon, Hydrogen, Nitrogen, Oxygen, Phosphate, and Sulphur). We have read how protocells might have originated containing molecules made of CHNOPS; a few could function as catalysts to carry out metabolic reactions and facilitate self-replication. However, all known life can arise only from pre-existing cells, which poses a pertinent question: how does the first cell originate from the protocell? To date, no scientist could create a living cell from organic molecules. The lack of a final demonstration of life from non-life leaves the question open, and research continues to fill in the knowledge gaps, showing that life's origin may be a gradual process across millions of years. Let me begin this chapter by narrating how science is piecing together the conditions and mechanisms by which the first cell might have arisen naturally, approaching this conundrum with evolving hypotheses and evidence.

## Floating signatures of life

Early chemical signatures were assumed to float as vesicles or coacervates in the' primordial soup'. Both forms of aggregates of organic biomolecules lack the double-layered lipid membranes characteristic of present-day living cells. During the emergence of life, vesicles formed from sheets of fatty molecules that curled into spherical shapes, encapsulating various biomolecules and organic polymers—essential components for life. On the other hand, coacervates were like droplets formed from the accumulation of organic polymers. These bubbles and droplets can be imagined as droplets of cooking oil suspended in water

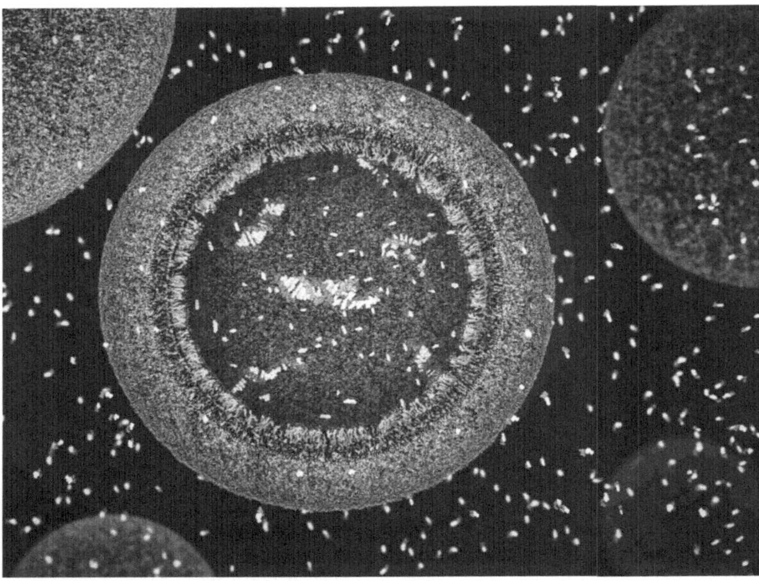

Figure 2.1: Illustration of a simple protocell consisting of a membrane boundary composed of fatty acids, with small pieces of primordial genetic material encapsulated in the internal space [*Image credit to Janet Iwasa; Courtesy Prof. Jack W Szostak, Harvard Medical School*]

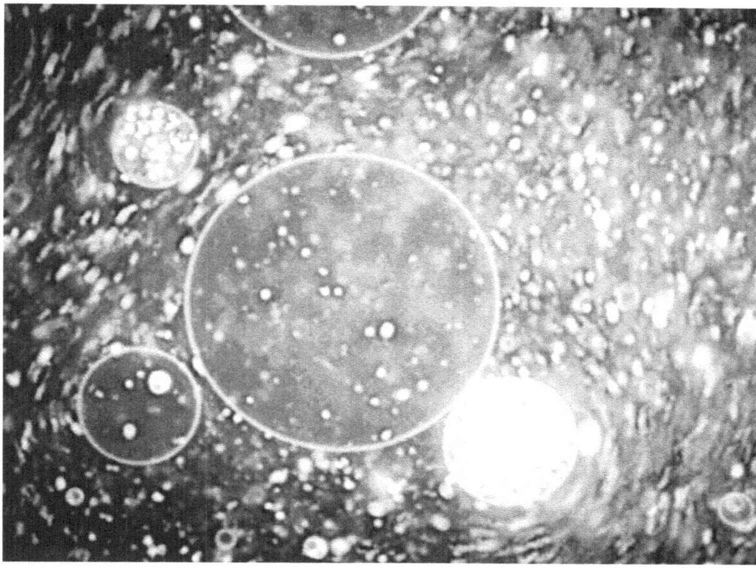

Figure 2.2: Fluorescence microscopic image of membrane vesicles composed of oleic acid/oleate [*Image credit to Martin Hanczyc; Courtesy Prof. Jack W Szostak, Harvard Medical School*]

During the journey to attain typical features of a cell, coacervates and vesicles underwent numerous rounds of self-assembly and fusion along with the internalisation of key molecules of life to get transformed into membrane-bound entities, which are referred to as protocells by origin-of-life researchers. Supporting evidence came from laboratory-made protocells (Figures 2.1 and.2.2]. Jack W. Szostak, Professor of Genetics at Harvard Medical School and winner of the 2009 Nobel Prize in Physiology and Medicine, is considered a pioneer in making model protocells since the early 2000s. Szostak and his team could successfully construct model protocells in the laboratory setting and have illustrated how these entities could absorb nutrients that supported their growth. These protocells, the initial life-like structures, likely consisted of three key molecular components: a membrane, an information-carrying molecule, and catalytic molecules, and are regarded as the precursors to modern cells. Therefore, understanding the geological, chemical, and environmental conditions present at the time of life's origin is crucial.

## LUCA and tree of life

Despite the advancements and exciting discoveries made by origin-of-life researchers, there exists a knowledge gap regarding the transitional period of protocells in the RNA world to complex cell-based living organisms possessing DNA, RNA, and Protein as life-propagating and life-supporting molecules. Presumably, at a certain stage in its development, DNA was incorporated into an RNA protocell to store genetic information and proteins to enhance catalytic efficiency. This marks the emergence of a life form that is the common ancestor of all contemporary life on Earth --- LUCA, which stands for the 'Last Universal Common Ancestor' of known organisms. In today's biological landscape, DNA serves as the main hereditary material and has maintained this role throughout the evolutionary history of life since LUCA up to modern organisms (Figure 2.3). This stability can probably be attributed to DNA's greater stability than RNA due to its structural features (deoxyribose of DNA having one oxygen less than the ribose sugar of RNA), which makes it more resilient and less reactive. Many empirical and laboratory-based simulation studies indicated that evolution progressed several phases, initially transitioning from an RNA world to a combination of RNA and protein, ultimately leading to the current DNA-RNA-protein-based living world.

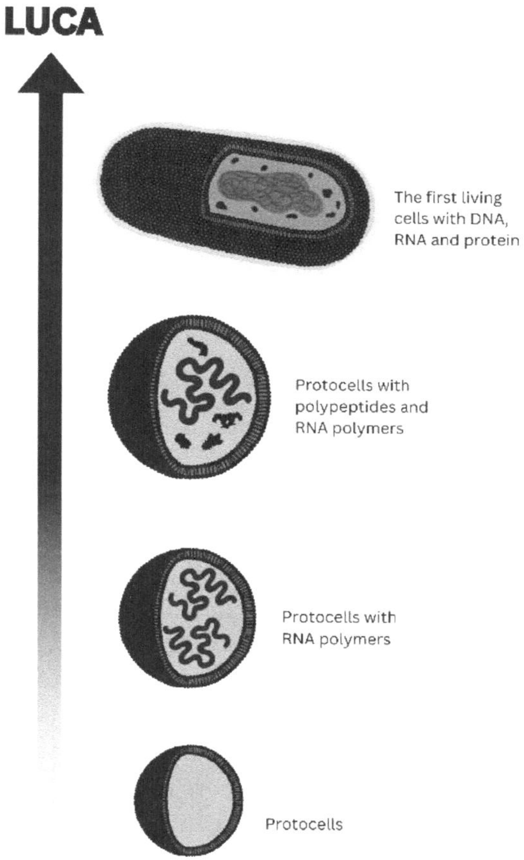


Figure : 2.3 Schematic showing hypothetical transitional path of protocells to LUCA [*Illustration credit: Uzma Shaikh, TIFR Mumbai, India*]

How many living organisms are currently present on Earth? According to estimates from the National Science Foundation's Tree of Life project, there could be anywhere between 5 million to 100 million distinct species inhabiting our planet. While scientists have documented nearly 2 million species on Earth, this estimate is believed to represent just a fraction of the total number of existing species today. An extensive fossil record exists for various plants and animals from the past, which may be ancestral relatives of modern-day organisms. Scientists have attempted to classify myriad species that are alive (extant species) and those that have vanished (extinct species) over time. Taxonomists who deal with the science of classification have established precise criteria to categorize these numerous species. Accordingly, six major kingdoms fall under three domains into which all organisms can be grouped.

## Domains of life

Taxonomy plays an essential role in classifying organisms based on three distinct criteria. Firstly, the organism's number of cells: unicellular (single-celled) or multicellular (many-celled). Secondly, the type of cells found in the organism is prokaryote (i.e., cell without a nucleus and internal organelles) or eukaryote (i.e., cell with a distinct nucleus with chromosomes containing DNA as genetic material). Thirdly, the mode of nutrition exhibited by the organisms: autotrophs (who can make their own food by photosynthesis, i.e. from solar energy) or heterotrophs (who can either feed on other organisms like a consumer or absorb breakdown materials from decaying dead organisms like a decomposer). While all prokaryotes are exclusively single-celled, eukaryotes can be single-celled or multicellular. Thus, organisms can be divided into three main domains: Archaea, Bacteria, and Eukarya. The LUCA, the hypothesized organism from which all life on Earth is believed to have descended, serves as the branching point on the tree of life from which the fundamental prokaryotic domains, Archaea and Bacteria, diverged (Figure 2.4).

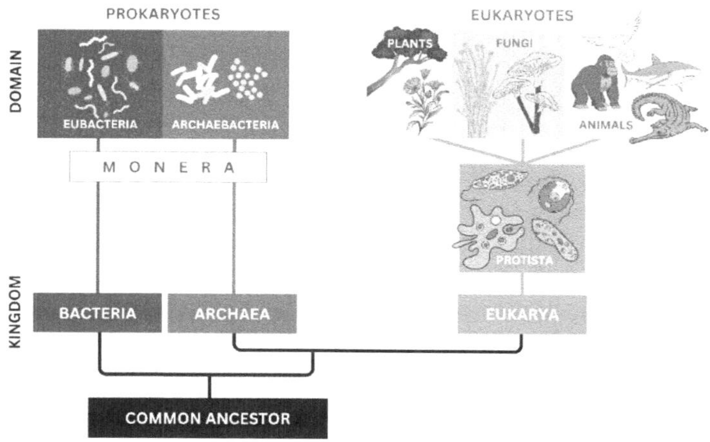


Figure 2.4 Tree of life with Domains and Kingdoms [*Illustration credit: Uzma Shaikh, TIFR Mumbai, India*]

Organisms within the Archaea domain are single-celled prokaryotes, often inhabiting extreme environments such as volcanic hot springs and brine pools and can thrive in places where oxygen is absent. Their cell walls do not contain peptidoglycan (unlike members of domain Bacteria); instead, they have unique lipids in their membranes that are absent in other life forms. Members of the domain Bacteria also consist of single-celled prokaryotes. Unlike the members of the domain Archaea, bacterial cells possess thick, rigid cell walls made primarily from peptidoglycan. The group exhibits significant ecological diversity- from free-living soil bacteria to harmful parasites – with some capable of photosynthesis, while others cannot survive in oxygen-rich environments. The third domain, Eukarya, encompasses all single and multicellular eukaryotes, i.e., organisms with a distinct nucleus inside their cell.

## Kingdoms of life

The three domains categorize the diverse range of living organisms into six kingdoms, as illustrated in Figure 2.4 Among the six kingdoms, two kingdoms are prokaryotes (also known as Monerans: Eubacteria and Archaebacteria) and are all single-celled, while the remaining four eukaryotic kingdoms (Protists, Plants, Fungi, and Animals) are single-celled or multicellular. The exact number of species currently on Earth remains uncertain. In 2022, Camilo Mora and her team estimated approximately 8.75 million living species (Reference: "How many species are there on Earth and in the ocean" - PLOS BIOLOGY

https://doi.org/10.1371/journal.pbio.1001127). However, less than 25% of these species could be documented to date. Countless undiscovered species inhabit our planet and are awaiting identification. Nevertheless, the overview of the six kingdoms of life summarised below shows how life on Earth portrays a stunning array of diversity.

**Kingdom Protista:** Most protists are unicellular eukaryotes. Some are autotrophic (i.e. able to prepare their own food) and can perform photosynthesis, while others are heterotrophic. Certain single-celled protists exhibit features reminiscent of fungi (e.g., Yeast), plants (e.g., Chlamydomonas), or animals (e.g., Amoeba).

**Kingdom Fungi:** Members of this kingdom are multicellular eukaryotic organisms. These heterotrophic organisms feed on dead or decaying organic matter by secreting digestive enzymes that break down food into smaller molecules for absorption.

**Kingdom Plantae:** Members of Plantae are predominantly multicellular autotrophs with cellulose-based cell walls and are non-motile. Plants are capable of synthesizing nutrients by sunlight (hence called photosynthesis) using the green pigment chlorophyll.

**Kingdom Animalia:** Members of Animalia are multicellular heterotrophs, lacking both cell walls and chloroplasts. Animals ingest food and digest it internally. Nerves, muscles, and specialized sense organs are unique in animals. Humans are biologically a part of the Kingdom Animalia and can be considered the most advanced form of animal living on Earth today.

## Any place for Virus in the tree of life?

We are all familiar with viruses, and even those without a scientific background can recognize the impact of a viral infection. The COVID-19 pandemic has heightened public awareness of viruses. However, viruses are often excluded from the overall classification scheme of life and living organisms. Viruses are often overlooked in evolutionary models because they are considered non-living entities. A virus is a sub-microscopic infectious agent whose origins are poorly understood, leading to ongoing debates about its classification. Viruses can infect all types of life, including animals, plants, and microorganisms like bacteria and archaea. Many view viruses as non-living entities that exploit living cells for replication. While viruses contain DNA and/or RNA, which could suffice to classify them as living simultaneously, they are also seen as non-living because they depend on a host organism for reproduction and energy. As a result, viruses exist in a grey area between living and non-living entities as they do not independently demonstrate growth or reproduction—characteristics crucial to be linked to life. Considering the conventional grammar of life, researchers typically consider viruses separately from other life forms while discussing the origins of life. To avoid diverting the book's focus, I will refrain from discussing viruses while highlighting the diversification of life across the geological timeline.

## Geological calendar and evolutionary clock of life

The geological processes have consistently occurred in various forms throughout the transition from the chemical evolution of life to biological evolution. Thus, I am directing your focus to the geological timeline of events in relation to the evolution and diversification of life, spanning from protocells to cells and from single-celled organisms to multicellular life forms. The geologic time scale catalogues Earth's history into four divisions: Pre-Cambrian, Palaeozoic, Mesozoic, and Cenozoic eras (Figure 2.5 and Table 2.1). To grasp the history of life, it is important to examine the geologic time scale, which is constructed from geological events, rock formations, shifts in the Earth's crust, the movement of landmasses, and the creation of present-day continents, oceans, and mountain ranges over hundreds of millions of years. Geologists, palaeontologists, and many other Earth scientists use the geological time scale to interpret the timing of some historical events on Earth. The time since the formation of the Earth has been divided into eras defined by significant changes over billions of years. The scale has different units; An Eon lasts over half a billion years. Eons are then divided into Eras that can last hundreds of millions of years. Eras are divided into Periods, again further divided into smaller units called Epochs. Distinct events define each era, like life's emergence, evolution, and extinctions.

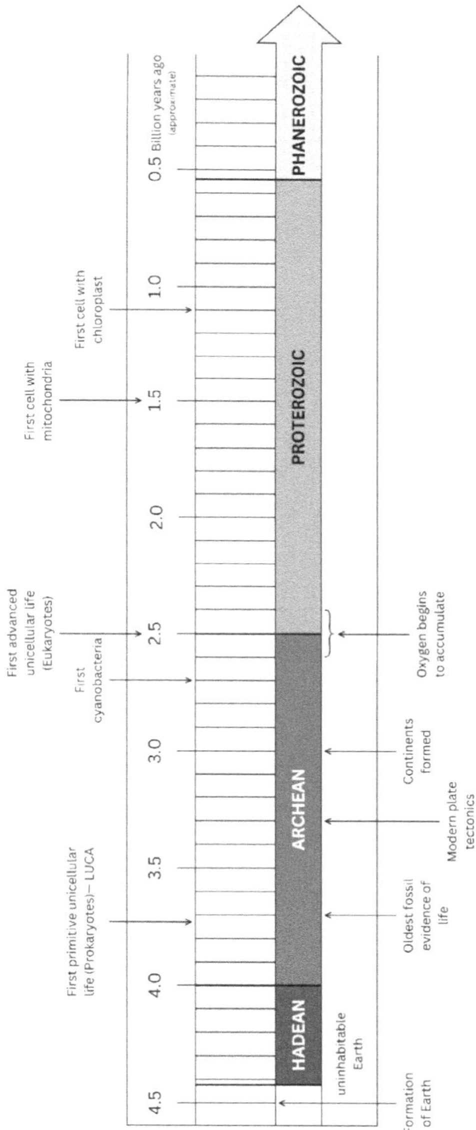


Figure: 2.5 Precambrian timeline and history of early life on Earth (4.5 billion to 542 million years ago). [*Illustration credit: Uzma Shaikh, TIFR Mumbai, India*]

Table 2.1: Geological Time Scale: Precambrian to recent.

| Eon | Era | Period | Epoch | Date (mya)* | Salient events |
|---|---|---|---|---|---|
| PHANEROZOIC | CENOZOIC | *Quaternary* | Holocene | Today → 0.01 | The last ice age ended about 12,000 years ago. By that time, our own species, the modern man - - *Homo sapiens*, had evolved and become dominant—extinction of large mammals in the northern hemisphere. |
| | | Age of Mammals | Pleistocene | → 1.8 | Earth's climate cooled, leading to a series of ice ages. Modern humans appeared |
| | | | Pliocene | → 5.3 | The first hominids (human ancestors) appeared. Many mammals increased in size and filled all the niches vacated by the dinosaurs. |
| | | *Tertiary* | Miocene | → 23 | Grass became widespread, and the Co-evolution of insects and flowering plants |
| | | | Oligocene | → 34 | Mammals dominant, animals inhabiting rainforests widespread, pigs, cats, rhinos, snails & bivalves (oysters) dominant in ocean |
| | | | Eocene | → 56 | Large mammals abundant, rodents, primitive whales appeared, and grasses appeared |
| | | | Paleocene | → 65 | Early mammals with placenta, first primates, modern birds, |
| | MESOZOICC | *Cretaceous* | | → 145 | Extinction of dinosaurs and many plants, birds appeared, origin of early marsupial mammals, ants, bees, butterflies, and flowering plants |
| | | *Jurassic* | Age of Reptiles | → 199 | Dinosaurs diversified, Seedless (gymnosperms) plants dominated on land, feathered dinosaurs and birds appeared |
| | | *Triassic* | | → 251 | Reptiles were dominant, First dinosaurs, the first mammals, and the first true flies appeared |
| | PALEOZOIC | *Permian* | Age of Amphibians | → 299 | Amphibians dominated, the largest mass extinction the Earth has ever encountered, and 50% of all animal families became extinct. |
| | | *Carboniferous* | Age of Fishes | → 359 | First reptiles and winged insects, mosses, ferns, and seedless vascular plants appeared. |
| | | *Devonian* | | → 416 | Diversification of jawed fish, early amphibians, many corals and many land plants |
| | | *Silurian* | | → 443 | Rise of vascular land plants, early jaw fishes, early insects, first spiders, scorpions, centipedes |
| | | *Ordovician* | Age of Invertebrates | → 488 | Diverse marine life, primitive fungi, first land plants, corals, molluscs, echinoderms, etc. |
| | | *Cambrian* | | → 542 | Rise of all major animal groups, Trilobites dominant, earliest chordates appeared |
| ***Proterozoic*** | 4.5 billion to 542 mya | | | | First, multicellular life, colonial algae, and soft-bodied invertebrate animals appeared, and the ozone layer formed in the atmosphere. |
| ***Archean*** | | | | | The origin of earliest unicellular life forms, LUCA, lived ~4200 my ago, photosynthesizing cyanobacteria appeared raising the oxygen level |
| ***Hadean*** | | | | | Cooling & Solidification of Earth's crust; No life forms known; atmosphere & ocean formed |
| *FORMATION OF EARTH* | | | | | |

*mya –million years ago

[Geochronological data from "The Geologic Time Scale 2012" Vol.1.Eds. Grandstein et. al. 2012 Elsevier]

## Microfossils: Signature of life on rocks

Pre-Cambrian Eons commenced 4.6 billion years ago with our lifeless planet. This time zone remained devoid of life for billions of years. Those who know about the existence of fossils (preserved remains, impressions, or remnants of a living entity before) would ask if anyone looked for 'microscopic' fossils dating back to the pre-Cambrian time. The answer is 'yes'. Prof. J. William Schopf, the director of UCLA's Centre for the Study of Evolution and the Origin of Life, obtained the rock containing microfossils from a deposit known as Apex chert in Western Australia in 1982. This site is one of the few locations on Earth where geological traces from early history have been preserved due to its lack of exposure to processes like burial or intense heat associated with tectonic activity.

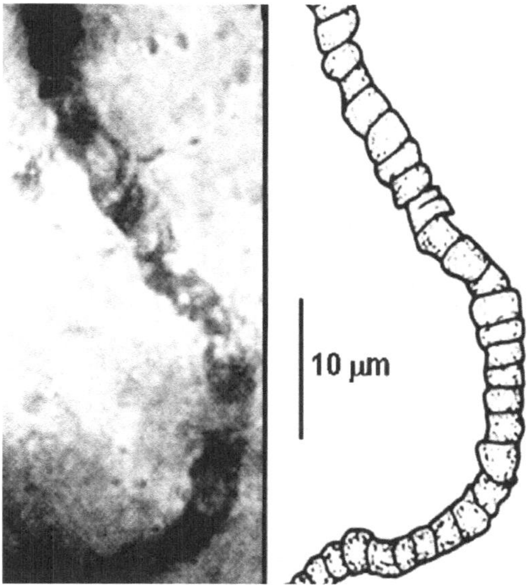


Figure: 2.5. A representative specimen of microfossil discovered in a sample of rock from the Apex chert deposit of Western Australia [Image courtesy J. William Schopf, Professor of Earth Sciences at the University of California Los Angeles]

An illustration (Figure 2.5) included here was one of the microscopic fossils discovered by Prof. Schopf, who published the work in the *Proceedings of the National Academy of Sciences (2017)* with John W. Valley, a geoscience professor at UW–Madison. They verified that microscopic fossils within a nearly 3.5-billion-year-old rock formation in Western Australia represent one of the oldest known fossils. Having considered the conditions of the Earth during pre-Cambrian times, these microbes exemplify the extremophilic lives of Archaea, adapted to thrive under extreme conditions.

Figure 2.6: A preserved sample of Archaean Stromatolite, the oldest (approximately 3.5 billion years old) living fossils on Earth sampled from Western Australia. Stromatolite rocks represent fossilised cyanobacteria that can be traced in each wavy layer (or mat). Each layer is formed by numerous colonies of cyanobacteria coalescing during fossilization. These ancient cyanobacteria grew in shallow sea water because they needed sunlight to photosynthesize.

The oldest well-studied fossils are cyanobacteria (earlier known as blue-green algae), found in Archaean rocks, popularly known as stromatolites coastal parts of Western Australia, dating back approximately 3.5 billion years. The

Stromatolites serve as contemporary representations of the earliest life forms identified on Earth. Their name, derived from the Greek meaning 'layered rock,' refers to microbial reefs formed by cyanobacteria. Modern stromatolites were discovered in Shark Bay, Australia, in 1956 and are present throughout various marine and non-marine settings across Western Australia.

## *The great oxidation event & Cambrian explosion*

Before the existence of cyanobacteria, Earth's atmosphere contained merely 1% oxygen. Photosynthesizing stromatolites for approximately two billion years thereafter released oxygen into ocean waters (akin to underwater trees before terrestrial trees emerged). Once these oceans became saturated with oxygen, they began to escape into the atmosphere, increasing atmospheric oxygen levels by around 20%, enabling life to thrive and evolve significantly. These congenial conditions of Earth led to the most rapid evolution of new animal species during the Cambrian period, popularly known as the "Cambrian explosion".

Figure: 2.7 One of the natural sites of Stromatolites, the ancient fossilized 'microbial mats', still preserved for billions of years. Cyanobacteria housed in these sites are now widely believed to have played a crucial role in transforming our atmosphere into one that is rich in oxygen, enabling the evolution of other forms of life [*Photo credit to Bush Heritage, Australia; Courtesy Ms Coco McGrath*)

Environmental transformations such as rising atmospheric oxygen levels and increased oceanic calcium concentrations may have contributed to the Cambrian Explosion. The existence of a continental shelf with various shallow pools may have created the essential habitats for larger animal species to thrive together. This phenomenon, characterized by the relatively swift emergence of most major animal phyla over a span of millions of years—approximately 530 million years ago—has been evidenced in the fossil record. There is a consensus among scientists that the Cambrian period represented an extraordinary era of rapid evolution and diversification of animal life, a pace of change unparalleled in the annals of history. Phanerozoic Eon started with the Cambrian Explosion, a pivotal event characterized by the extensive diversification of multicellular organisms.

## Visible footprints of past life

The Phanerozoic eon, spanning from 541 million years to the present, is characterized by its name derived from the Greek words *'phanerós'* and *'zōḗ'*, meaning 'visible life.' Further, the term Phanerozoic reflects the extensive fossil records found within its rock formations. Organisms that thrived during this eon possessed hard body structures, such as claws, scales, shells, and bones, facilitating their preservation as fossils. The Phanerozoic eon is divided into three distinct eras: Paleozoic ("ancient era"), Mesozoic ("middle life"), and Cenozoic ("recent life).

The Phanerozoic Eon witnessed the emergence of various life forms: 'the age of invertebrates, 'the age of fish'; 'the age of amphibians, 'age of reptiles, the emergence of birds, and 'the age of mammals' including modern man. During this era, various animal and plant species have thrived, diversified, and established themselves in numerous habitats across the Earth's surface. Five significant and abrupt declines in biodiversity, known as mass extinctions, have taken place at five distinct intervals during the Phanerozoic eon, resulting in the rapid loss of numerous taxa over a brief geological timeframe, which will be the primary focus of our upcoming chapter (Chapter 3).

# CHAPTER 3

# *Lost World of Life*

Don't you think that the world we inhabit now was quite different 250 million years ago? However, how different can one truly imagine it to be? While writing this chapter, I remember meeting a geologist and a research student from the Wadia Institute of Himalayan Geology during a trek in the Himalayas. They presented me with fascinating fossils of marine creatures discovered in the Himalayas. It was quite challenging for me to conceive that these mountains were once submerged beneath the sea. The geologist invited me to his library, where he shared several maps which he used for teaching. These maps felt like a time machine, transporting me back to an Earth 250 million years ago, reminiscent of a science fiction narrative. If an author had crafted a geography book and created a map depicting that era, it would appear vastly different from the maps found in contemporary geography textbooks used in schools. My geologist companion illustrated various geological phases and elucidated how the Earth has continuously evolved over millions of years, reshaping land and oceanic landscapes. Landmasses that were once connected eventually drifted apart, each moving away from the other. Finally, my fantasy was over, and I returned to the present day after a virtual tour of the primitive earth through this metaphorical time machine. My mind could resolve Earth's remarkable landmass transformation while reflecting on the fate of living organisms. This chapter opens with an account of Earth's extraordinary changes before exploring what occurred to living beings during this period. Did they all perish without a trace? That question encapsulates the essence of this chapter.

## Ancient geography of the Earth

Today, the landmass of Earth is categorized into seven continents: Asia, Africa, North America, South America, Antarctica, Europe, and Australia. However, these continents were once unified as a vast supercontinent known as Pangaea, which translates from Greek as "all land." This landmass was encircled by a singular ocean termed Panthalassa, meaning "all sea" in Greek. Over a century ago, German scientist Alfred Wegener introduced the concept of Pangaea and the 'continental drift' theory, which aimed to explain how this supercontinent fragmented and shifted over time (Figure 3.1). The 'plate tectonics' theory has superseded the earlier notion of continental drift.

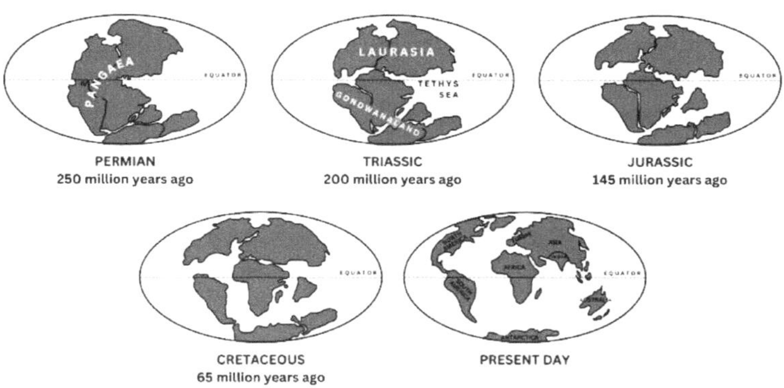


Figure 3.1. The supercontinent Pangea: The geological periods indicating when it existed and how it ended breaking apart and forming modern continents. (Illustration credit: Uzma Shaikh, TIFR Mumbai India)

A tectonic plate is a massive, irregularly shaped slab of solid rock varying from hundreds to thousands of kilometers.

These disjointed plates form Earth's upper crust and glide atop the mantle, our planet's extremely hot interior layer. The motion of these tectonic plates is driven by convection currents within the mantle, which circulate molten materials through rising and sinking motions. Such movements can result in earthquakes, volcanic eruptions, and other significant environmental events. To visualize this, consider the vibrations on the lid of a boiling pot; similarly, the intense heat and pressure within Earth's interior instigate the movement of tectonic plates. These plates interact in several ways: collide, pull apart, or slide past each other. The action of tectonic forces can result in the formation of mountain ranges, such as the Himalayas.

## Splitting of continents

It is suggested that shifts caused by tectonic forces rearranged both oceanic and continental crusts and significantly impacted Earth's biodiversity. Scientists have compiled evidence outlining a world map from the Paleozoic era (approximately 542 million years ago), highlighting 'Gondwana' as Earth's largest landmass for over 200 million years. The eventual collision of various land masses led to the formation of Pangaea (around 300 million years ago). Approximately 200 million years ago, Pangaea began to fracture due to extensive tectonic activity into two principal regions: Laurasia (the northern part) and Gondwana Land (the southern section). Around 135 million years ago, Laurasia continued its movement and fragmented into what are now North America, Europe, and Asia. Meanwhile, Gondwana Land gradually separated into present-day continents, including South America, Africa, Australia, and the Indian subcontinent (Figure 3.1).

## Fate of Gondwana-Pangean life

The surface of Earth has consistently changed; its volatile crust and inner core have periodically transformed oceanic and continental landscapes. Life on Earth has always faced pressures that challenge its survival or lead to extinction. Tectonic activities give rise to drifting continents and mountain formations that isolate species previously cohabiting in an area. Dynamic shifts among tectonic plates established distinct 'realms,' creating biogeographical divisions across Earth's surface and specific 'endemic' zones where certain plants or animals are confined to their habitats without being found elsewhere (e.g., kangaroos exclusive to Australia). Geological processes during periods marked by Gondwana-Pangea resulted in 'ecosystem evolution' while presenting challenges for organisms striving to adapt or face extinction—a topic explored further in this chapter. Significant ecological legacies from Gondwana-Pangea were disrupted due to large-scale tectonic activities around 200 million years ago when the Atlantic Ocean formed for the first time alongside splitting landmasses that separated South America from Africa—where they remain today. This geographical isolation facilitated unique rainforest ecosystems around equatorial regions on Earth; flora and fauna adapted here became endemic while others faced extinction over time. Unique ecosystems evolved independently across South America, Australia, Africa, and other post-Pangean areas devoid of competition from species originating on separate continents.

## Extinction: Dying out of a species

The process through which a new species arises is termed 'speciation.' However, if the last surviving individual of a species passes away and there are no remaining members capable of reproduction, the species is considered extinct.

Extinction and speciation are opposing mechanisms that influence the evolution of life on our planet. When speciation rates surpass extinction rates, it increases biodiversity—the variety of living organisms. Conversely, when extinction outpaces speciation, biological diversity diminishes. Generally, there exists a balance between these two processes; however, there have been five significant events in Earth's history known as 'mass extinctions,' which will be explored later in this chapter. Scientists estimate that around 99.9% of all plant and animal species that ever existed—approximately five billion distinct forms—are believed to have disappeared from Earth. Over millennia, various species have emerged and vanished as they adapted to changing environments. Each extinction event, gradual or abrupt, has played a role in shaping the complex biological history of our planet by paving the way for new species to develop and evolve.

While human activities have undeniably accelerated the extinction rates of numerous species, it is important to recognize that natural phenomena primarily caused the Big Five mass extinctions. The underlying causes of these historical extinction events remain elusive, compelling scientists to rely on geological evidence from rock formations and fossil records—often incomplete. Throughout geological history, many organisms have gone extinct due to their inability to adapt and survive under various natural conditions that have been documented and studied. Some key factors include:

*Volcanic Eruptions:* Volcanic activity was a significant natural event during the Precambrian era. The release of hot molten lava and volcanic ash that created a cloudy atmosphere blocked sunlight for extended periods, rendering conditions harsh for life forms. Volcanism was notably prevalent during periods of plate tectonic movement.

*Climate Change: Global Cooling and Ice Ages:* The current Cenozoic era (66 million years ago to present) has witnessed several ice ages as Earth experienced extensive freezing events with glaciers advancing from mountains onto land and ocean surfaces. Researchers believe at least five major ice ages occurred: Huronian (2.4-2.1 billion years ago), Cryogenian (850-635 million years ago), Andean-Saharan (460-430 million years ago), Karoo (360-260 million years ago), and Late Cenozoic-Quaternary (34 million years ago-present).

*Impact from Near Earth Objects:* Earth is constantly bombarded by objects from outer space, known as Near Earth Objects (NEOs), which include asteroids, comets, and meteorites (fragments of asteroids that become visible as burning 'meteors' upon entering Earth's atmosphere—commonly referred to as 'shooting stars' or 'falling stars'). Evidence suggests that numerous impacts have occurred throughout Earth's history. Large objects, such as asteroids, have led to catastrophic consequences, wiping out entire populations of various species.

*Hypoxic Conditions in the Sea:* Research into Earth's history has revealed that extinction events have been linked to ocean waters warming and losing their dissolved oxygen (resulting in hypoxic conditions), essential for marine organisms' respiration and survival.

*Eco-Biological Stress and Non-Climatic Factors:* Excluding the major evolutionary catastrophes, background extinction occurs continuously and naturally. Various events, including habitat loss due to climatic changes, natural disasters, and the introduction of invasive species, can precipitate population declines. Increased predation, disease spread from new pathogens, and competition for limited resources can intensify these declines. Additionally, factors such as reduced genetic diversity, lack of fitness due to prolonged inbreeding in a small-sized population, infertility, and various reproductive failures and challenges contribute to species' eventual extinction.

## Mass extinctions saga: The 'Big Five'

Throughout Earth's history, there have been five significant mass extinctions, commonly called the "Big Five." These events were characterized by a rate of extinction that far exceeded the typical background rate. Scientists define a mass extinction event as one in which at least 75% of species are lost within a span of less than two million years—a relatively brief period from a geological perspective. Figure 3.2 illustrates the timeline showcasing these five major extinction episodes marked by sudden spikes in extinction rates.

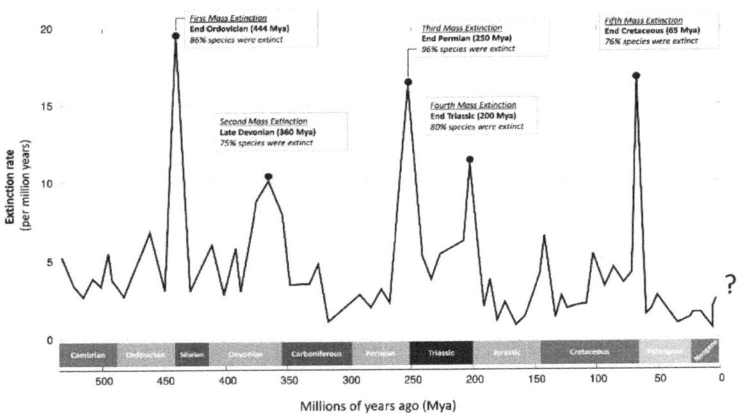


Figure 3.2. Graphical representation of 5 episodes of mass extinctions as reflected in the fossil record. [adapted from the source data at https://ourworldindata.org authored by Hannah Ritchie] (Image credit: Dr Vikas Kashid, MIE-SPPU IoHE, Doha Qatar)

*The 'First Big' -- The first jolt for evolution of life:* Around 444 million years ago, the planet experienced its first major mass extinction, eradicating nearly 86% of all species. This event was primarily driven by continental drift and extreme climatic fluctuations. Initially, an ice age created harsh conditions, followed by a dramatic rise in ocean levels as the ice melted. The tectonic uplift of mountain ranges played a dual role: It destroyed existing habitats and forged new environments that would eventually support different life forms.

*The 'Second Big' –The Devonian tragedy*: Approximately 360 million years ago, during the Devonian period, the second great extinction occurred, claiming about 75% of species. Factors contributing to this crisis included hypoxic ocean conditions, intensified volcanic activity, and a rapid decline in global temperatures. This combination of elements created an environment inhospitable to many marine organisms, leading to widespread die-offs (Figure 3.3).

Figure. 3.3. Photo of a fossil of a fish from Devonian period (approximately 400 mya)

*The 'Third Big' – The 'Great Dying'*: The Permian period, which ended roughly 250 million years ago, witnessed the most severe extinction event known, often called the 'Great

Dying'. An astounding 96% of species vanished, driven by catastrophic geological events. Massive volcanic eruptions in present-day Siberia released unprecedented lava and greenhouse gases, triggering global warming and prolonged acid rain. The scale of these eruptions was staggering, with estimates suggesting over 3 million cubic kilometres of lava being expelled alongside 14.5 trillion tons of carbon compounds. This massive eruption that formed the traps (known as 'Siberian Traps) is one of the largest volcanic events in the last 500 million years. These changes precipitated a cascade of environmental transformations, particularly affecting marine ecosystems and leading to the extinction of many arthropod groups (invertebrates with jointed leg and body cover made of water-impervious chitin) got wiped out including the trilobites, which were the dominant group on land (Figure 3.4).

Figure 3.4. Photo showing fossil remains of Trilobite in stone (collected from Elkai and Errani, Morocco). All trilobites died out by 251 million years ago.

*'Fourth Big' – The tectonic terror on Earth*: The fourth mass extinction transpired around 200 million years ago after the Triassic-Jurassic period and about 80% of species perished. The breakup of Pangea (Figure 3.1) initiated dramatic shifts in oceanic and atmospheric circulation patterns. Escalated tectonic activity led to extensive volcanic eruptions in what is known today as the Central Atlantic Magmatic Province (CAMP). Emissions of methane, sulphur dioxide, and carbon dioxide induced greenhouse effects that drove global warming and elevated sea levels, among other consequences. High carbon dioxide concentrations also contributed to ocean acidification; marine organisms with shells struggled to form due to acidic conditions that dissolved calcium carbonate-containing shells. Despite the high extinction rates among flora and fauna, the earliest dinosaurs survived and rapidly diversified during the Jurassic era (figure 3.5). While the oldest dinosaurs of the Triassic were relatively small, they adapted to various ecological niches left vacant by less resilient reptiles. The mechanisms behind their survival during this mass extinction remain poorly understood, but they eventually evolved into diverse forms, marking the onset of the 'age of reptiles' (Figures 3.6; 3.7)

Figure. 3.5. Photo shows a life-size model of Tyrannosaurus rex skeleton (Source: Jurassic Park in Universal Studios, Singapore).

*The 'Fifth Big' – A farewell to dinosaurs*: The fifth period of mass extinction happened around 65 million years ago, and 76% of terrestrial and aquatic species went extinct. This was when all non-avian (without features of bird) Dinosaurs were wiped out from the face of the Earth. The fifth mass extinction is called the K-T extinction because it happened at the end of the Cretaceous period and towards the beginning of the Tertiary period. Why is this not called C-T? Geologists use 'K' as shorthand for Cretaceous to avoid confusion with the earlier Cambrian period. Alongside the dinosaurs, two other dominant groups, the mosasaurs (large marine reptiles) and pterosaurs (flying reptiles), also faced extinction. Numerous other species vanished, including ammonites (spirally shelled marine molluscs).

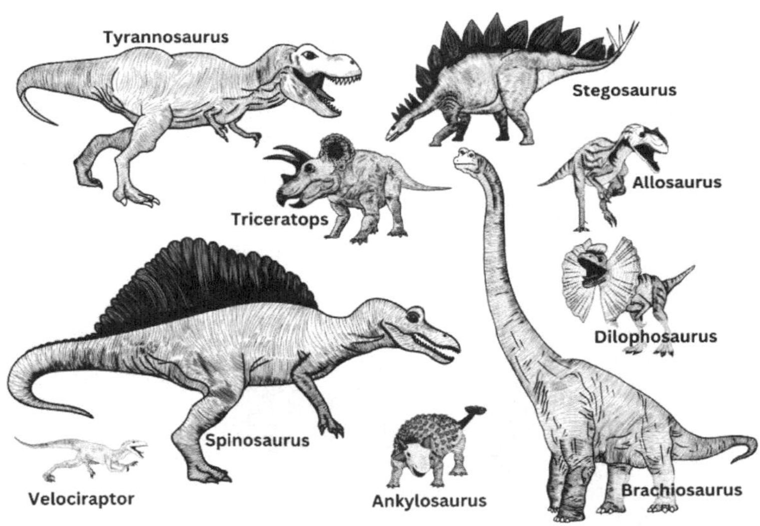


Figure 3.6. Schematic illustration showing major groups of dinosaurs (image not to scale). [Illustration credit: Uzma Shaikh, TIFR, Mumbai, India]

Figure 3.7. Photo of a fossilized dinosaur egg recovered from the nest site of dinosaurs, preserved in late Cretaceous Lameta sediment. The specimen is located at Dinosaur (*Ashmadha*) Fossils National Park, Kakra Kho, near Mandu, Dhar district, Madhya Pradesh, India. [Image credit: Anurita Nath]

Figure 3.8. (a) Photo of a fossil of *Archaeopteryx*, the bird-like dinosaur, discovered from (b) a stone quarry in Solnhofen, Bavaria, Germany. *Archeopteryx* lived around 150 million years ago.

## 'Fifth Big' survivors

Mammals emerged during the Triassic period, but they remained relatively inconspicuous until the extinction of dinosaurs at the end of the Cretaceous. The evolution of birds began in the Jurassic period, with the earliest birds evolving from a group of theropod dinosaurs. *Archaeopteryx* is recognized as the earliest known bird. Some consider it a bird, while others view it as a dinosaur. The fossilized skeleton of *Archaeopteryx* looks like that of a dinosaur, and it has teeth, whereas birds do not. It also had feathers modified for flight (Figure 3.8). *Archaeopteryx* is significant for illustrating the evolutionary link between birds and dinosaurs, serving as an intermediate fossil. Only a handful of ancestral bird lineages survived the mass extinction events and subsequently thrived. After the fifth mass extinction, Earth transitioned from the 'age of dinosaurs to the 'age of birds and mammals. Avian dinosaurs, which are bird-like, managed to survive and have given rise to modern birds, which have thrived. Species that endured the extinction were generally small and inhabited aquatic environments distant from the disaster sites or adapted to burrowing underground. Certain extremophilic insect groups survived the extinction and rapidly diversified by taking advantage of the new ecological opportunities that emerged following the biodiversity crisis. Many of their taxonomically related species exist today and diversified before and after the fifth mass extinction event (Figures 3.9, 3.10; 3.11). The fossil history of many arthropods (Figure 3.12) showed their presence from the early Paleozoic era to the present. The mammals present during the fifth mass extinction were also relatively small and swiftly adapted to occupy the niches left vacant by dinosaurs.

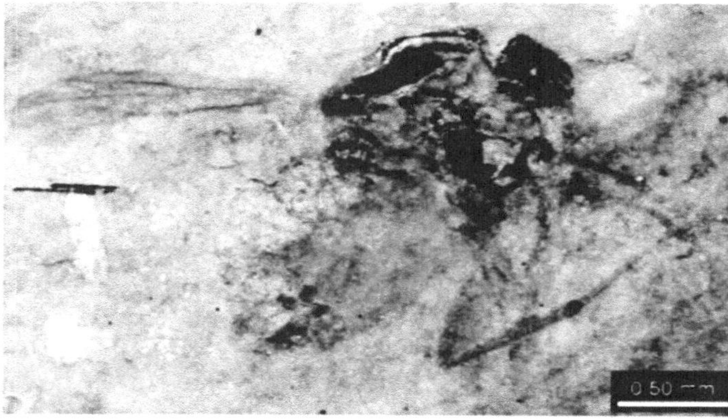

Figure 3.9 The photo shows the fossil imprint of a chironomid midge (*Podonomius blepharis* Lukashevich et Przhiboro 2011; doi:10.3897/zookeys.130.1555) in a stone collected from Mongolia and dated to around 150 million years ago. Photo credit and courtesy of Dr Elena Lukashevich (Borissiak Palaeontological Institute, Moscow, Russia) and Dr Andrey Przhiboro (Zoological Institute, St. Petersburg, Russia).

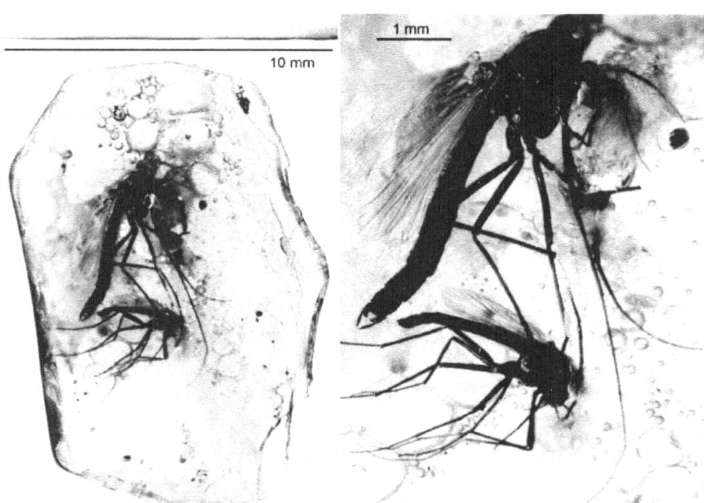


Figure 3.10 Photographs showing fossil of 'mime midges' (40 million years ago) preserved as inclusion in Eocene Baltic amber (Image courtesy: Prof. Wojciech Gilka, University of Gdańsk, Poland)

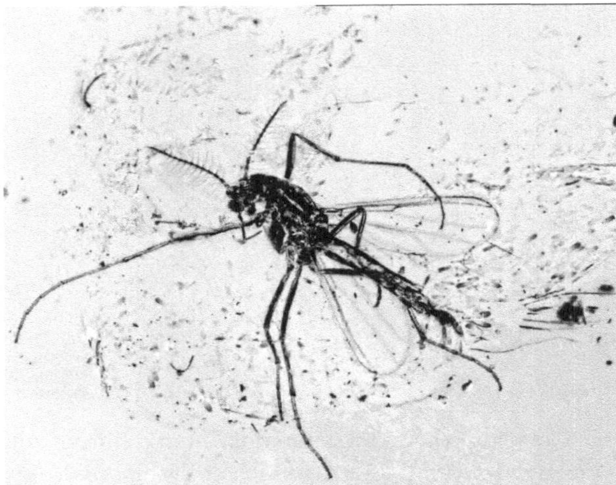

Figure 3.11 The photo shows a fossil of an adult insect (Chironomini male, cf. Xestochironnomus) in Chiapas amber, collected from the Simojovel region of Chiapas, Mexico, aged 15-23 million years ago. Material from the collection of Palaeo Evo Devo Group, University of Munich, Germany. (Photo credit and courtesy of Dr Viktor Baranov, Doñana Biological Station – CSIC, Seville, Spain).

Figure 3.12 The photo depicts a crab fossil (Arthropoda: Decapod) from the Cenozoic era. , [Photo credit: MUSE-Museo delle Scienze, Trento, Italy; Personal communication and courtesy Dr Valeria Lencioni]

## Final days of dinosaurs: A tale of dual catastrophe

Taking a trip back to the Triassic, Jurassic, and Cretaceous periods, we come across the magnificent group of reptiles --- the dinosaurs that once roamed, dominated, and ruled the Earth. The term 'Dinosauria' was first introduced in 1842 by Sir Richard Owen, a British naturalist. It comes from the Greek words: 'deinos,' which means fearfully great, and 'sauros,' meaning lizard. Dinosaurs were truly colossal creatures; some could grow up to 40 meters long and stand as tall as 18 meters! No other extinct species has captured our imagination like dinosaurs have. Consider classic literature like Sir Arthur Conan Doyle's *The Lost World* or hit movies such as Stephen Spielberg's *Jurassic Park*. Thanks to these films and TV shows, even kids today can easily recognize the mighty *Tyrannosaurus rex*—one of the most iconic members of this ancient group.

The Old School of evolutionary biologists believed that dinosaurs fell victim to competition from emerging small mammals and birds. Despite reigning for approximately 170 million years (Nature Communications 2021:12, 3833), their total disappearance from the planet was abrupt (according to fossil records). Over time, numerous theories have emerged regarding their demise; however, two main scientific explanations have gained considerable attention: the extra-planetary impact theory and the Deccan Traps volcanic eruption theory. It is a tale of two catastrophes that changed the face of reptilian life on Earth.

### *Extra-terrestrial impact theory*

In 1979, geologist Walter Alvarez and his father, physicist Luis Alvarez, a Nobel Prize laureate, and their team identified a layer of iridium at the KT boundary near Gubbio,

Italy. This layer dates back to the Cretaceous (K) and Tertiary (T) periods. Iridium is uncommon in the Earth's crust but is found in greater concentrations in the Earth's core, asteroids, and comets. Their findings, published in 1980 and referred to as the Alvarez hypothesis, suggested that an asteroid impact on Earth approximately 65 million years ago could have been responsible for the mass extinction event at the end of the Cretaceous period.

## Impact of asteroid— The Chicxulub 'smoking gun'

Discovery of the Chicxulub crater on Mexico's Yucatan Peninsula supported Alvarez's theory regarding a significant impact contributing to Earth's fifth mass extinction. It is now widely accepted that an asteroid approximately 10 kilometers (6 miles) wide struck Earth at Chicxulub around 65 million years ago. This crater ranks among the largest on our planet, measuring about 180 kilometers (112 miles) in diameter and reaching depths of up to 900 meters (3,000 feet). The collision generated immense tsunamis and ignited wildfires across vast areas (Figure 3.12). A massive column of gas and dust likely burnt all living organisms. The impact acted as a 'smoking gun,' propelling enormous amounts of particulate matter into the atmosphere, which shrouded sunlight for years, leading to a collapse of photosynthesis. As a result, herbivorous species, including many dinosaurs, faced starvation. The energy released from this impact was equivalent to roughly 100 million megatons of TNT. Accompanying phenomena included wildfires, acid rain, ocean acidification, and severe global warming—factors that collectively eliminated non-avian dinosaurs and other species unable to adapt to these drastic changes. Most of the scientific community now acknowledges Chicxulub as the 'smoking gun' that contributed to the extinction of over 75% of living species on Earth.

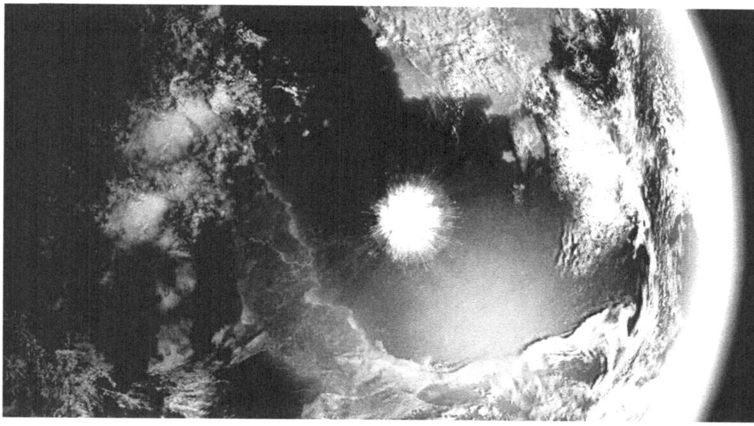

Figure 3.12 Illustration of a large asteroid colliding with Earth on the Yucatan Peninsula in Mexico.

## Impact of volcanism— the Deccan Traps

During the late Cretaceous period in India, extensive volcanic eruptions produced vast lava flows observable at the K-T boundary; these geological formations are known as the Deccan traps, which appear like ruptures in the Earth's surface. They get the name "traps" due to the step-like hills that arose from those lava flows. The term 'trap' comes from the Swedish word 'trappa', meaning steps, as the landscape of the hill looks like steps in stairs (Figure 3.13). The term "Deccan" is derived from a Sanskrit word meaning "south," which points to their location in southern India. Interestingly, these are not the usual cone-shaped volcanoes present elsewhere. Here, it is something geologists refer to as "flood basalt", meaning they came into being when gigantic amounts of lava burst through the cracks in the Earth's surface, spreading out lava like a river made of molten rock. Analysis reveals that these lava rocks originated from deep within Earth's mantle—a region also relatively rich in iridium—

challenging Alvarez's assertion that volcanism alone could account for this rare element's presence on Earth's surface.

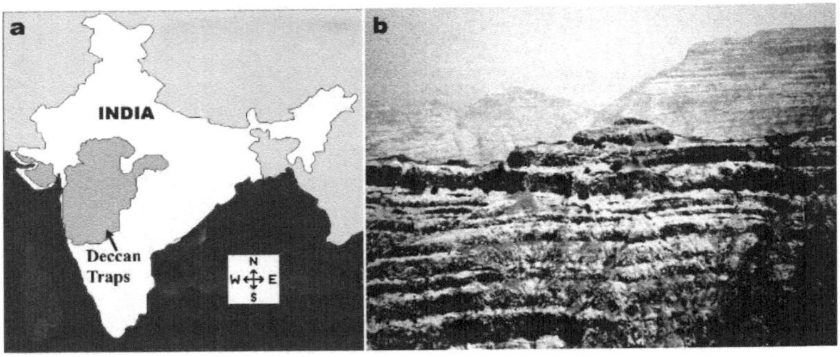


Figure 3.13 (a) Schematic map of India showing the area of 'Deccan Traps' (b) view of 'Deccan traps' showing stacks of historical lava flow (the horizontal layers in rocks seen in hills) in the Western Ghat region, near Mahabaleshwar, India. [Image credit: Anuraj Nath]

Earth scientists estimate the formation of the Deccan Traps began approximately 66.25 million years ago, coinciding with the end of the Cretaceous period, with the most intense eruptions occurring around 66 million years ago in the Western Ghats. This series of eruptions lasted less than 30,000 years and resulted in lava flows extending over 1000 kilometers across India, marking the longest flows known on Earth. The Deccan Traps contributed to a massive release of dust, ash, and sulfuric aerosols into the atmosphere, which limited sunlight and disrupted photosynthesis in plants. Furthermore, the increased carbon dioxide levels contributed to a rise in global temperatures, creating conditions sufficient to explain the fifth mass extinction. Interestingly, dinosaur eggshells, nests, and bone fragments are frequently found throughout central and western India.

## What killed dinosaurs?

Both catastrophic events—the extensive volcanism associated with the Deccan Traps in Western India and the massive asteroid strike at Chicxulub—coincided closely with geological timelines, impacting Earth dramatically. The aftermaths bore striking resemblances: both scenarios enveloped the planet in dust clouds and toxic gases while initiating wildfires alongside acid rains; they altered atmospheric chemistry as well as oceanic conditions, resulting in rising temperatures lethal for numerous species incapable of adaptation—including non-avian dinosaurs themselves. However, determining whether the Deccan Trap volcanism or the asteroid impact was solely responsible for the complete extinction of non-avian dinosaurs remains challenging. Both theories are supported by substantial scientific evidence, and it is generally believed that the combination of these two catastrophic events led to the extinction of dinosaurs and many other organisms.

A recent study by Dr. Alexander A. Cox and Dr. C. Brenhin Keller rekindled the debate, asserting that volcanism was likely the primary driver of the K-T mass extinction, placing the asteroid impact in a secondary role. While the evidence supporting the asteroid impact and volcanic activity is substantial, the debate continues over which events played a more pivotal role in the extinction of non-avian dinosaurs. Earlier studies, including one published by Prof. Paul R. Renne and colleagues in 2015, suggested that the Chicxulub impact may have triggered volcanic activity in India. There are many takers of this view, and now, many scientists have second thoughts about the possible synergistic effect between the two catastrophic events. In the words of Robert Sanders, "Asteroid impact and volcanism was a one-two punch for dinosaurs." This ongoing discourse highlights the complexity of the extinction event. It emphasizes the need for continued research into the relative contributions of Chicxulub impact and Deccan traps volcanic activity in understanding one of Earth's most defining moments.

While significant progress has been made in identifying the causes of the K-T mass extinction, the interplay between asteroid impact and volcanic activity remains a topic of scientific inquiry. As new data emerges, the narrative of the dinosaurs' demise continues to evolve, inviting further exploration into the intricate dynamics of our planet's history.

# CHAPTER 4

# *Endangered Life*

In 2019, the Intergovernmental Science-Policy Platform on Biodiversity and Ecosystem Services (IPBES) raised the alarm about nearly 1 million species of plants and animals facing extinction threats—more than at any other time in human history. So, are we actually living in a world that's at risk? The story of life on Earth is still being written, and this chapter will explore fresh perspectives, including those natural drivers of continuously present background extinctions and a few historical lesser-known events before humans even appeared on the planet.

## Is it all calm after the storm?

Is life on Earth truly without any extinction-level risk since the dinosaurs vanished? Whether the Earth has achieved a relatively stable state, in contrast to the frequent large-scale volcanic and tectonic activities that characterized its past several hundred million years, remains debatable. Since the fifth mass extinction approximately 65 million years ago, no significant extinction events have been comparable to the 'big five.' So, does that mean Earth is now a friendlier place for life and free from serious threats? Many of us would probably say, "Not really." Can we completely prevent extinction? And if not, what factors are quietly at play behind the scenes, contributing to this continuous basal extinction level and putting life on Earth under constant threat?

## Adapt or perish: the eternal law of life

Life on Earth projects a complex web of living relationships, constant challenges and eco-biological stress operating behind the scenes. Evolution is environment-driven, and life evolves based on its surroundings, which creates a "struggle for existence." Charles Darwin highlighted this essence of life and made us realise who makes it through and how the "fittest" among them keep reproducing to carry on their species. Those who can adapt to their surroundings are the ones who continue to pass on their genes, creating a legacy that lasts through generations. It's a bit like a game of musical chairs, where the environment (*nature*) plays the music and decides (selects) who gets to stay in the game and who gets left out.

## Competition: Live or die

One of the most important biotic environmental factors is competition. This competition can occur between individuals of the same species (intraspecific) or individuals of different species (interspecific). Competition for resources happens constantly in nature and has persisted throughout the evolutionary timeline. This competition frequently results in extinction events. Species not as well-adapted may have limited access to essential resources shared by both competitors. Consequently, individuals of the less adapted species face a lower likelihood of survival, which can ultimately lead to their extinction.

## Fear of aliens: Threat of invasive species

The invasive species infiltrate ecosystems that are not their original habitats and spread beyond their native geographical boundaries. They threaten the native species, the original inhabitants of the ecosystem. The invasive species reach new regions through natural migration, but they are typically introduced by the actions of other organisms (currently mostly by humans after their origin on Earth). Increasing temperatures and alterations in precipitation patterns can compel certain invasive species to expand into unfamiliar territories. Various invasive organisms, including insects, pathogens, and weeds, are capable of being carried by wind currents. Similarly, many aquatic animals can passively reach a new region during flooding or forced migration. Additionally, many pathogens and parasites accompany their hosts as they invade new environments.

Throughout Earth's history, species have migrated and established themselves in new locales. In some instances, migratory species failed to form viable populations in these new habitats and quickly became extinct. However, there were cases where some invasive species adapted and thrived, often altering the existing food web, outcompeting native species, and capturing vital resources such as food and habitat space. If the original native species cannot compete effectively, they may relocate and leave their own habitat or face extinction. The introduction of invasive species can have catastrophic implications for native flora and fauna. As these newcomers adapt and proliferate, they often disrupt established relationships within the ecosystem. Native species, unable to compete with the invaders, may be forced to relocate or face extinction.

Invasive species breed profusely in places far away from their original ecosystem, occupy more areas and soon alter their new territories. They can be categorized into various types, including predators, competitors, parasites, and pathogens that cause diseases in both plants and animals

## Invasion by alien predators: Kill or get killed

The immediate dangers posed by invasive predator species involve hunting down native animals and overshadowing them in competition, as these locals aren't equipped with any evolved defences to handle the threat. According to a report from 2016 by Tim Doherty and colleagues (Proceedings of the National Academy of Sciences, 113:11261-11265 DOI:10.1073/pnas.1602480113), thirty species of invasive predators are associated with the extinction or risk of extinction for 738 vertebrate species, accounting for 58% of all extinctions among birds, mammals, and reptiles. Notably, predatory feral cats and red foxes have driven two-thirds of Australia's digging mammal species to extinction over the last two centuries.

## Unwanted guest: The killer snake in Guam

Anyone would be shaken to hear about a major extinction event caused by predation in the wild on an island. One such case occurred on Guam, a little island in the western Pacific. This island is home to unique species that can't be found anywhere else. There were no natural predators, especially not for the birds living in the forests. The island's avian fauna faced a major shake-up when an invader appeared: the brown tree snake (*Boiga irregularis*).

After Japan invaded Guam following the Pearl Harbor attack in December 1941, Allied forces took it back by August 1944, turning it into a key base for bombing missions against Japan towards the war's end. During this turbulent wartime, it was believed that brown tree snakes—originally from the Solomon Islands and Papua New Guinea—managed to infiltrate military transport ships or aircraft arriving in Guam (Figure 4.1).

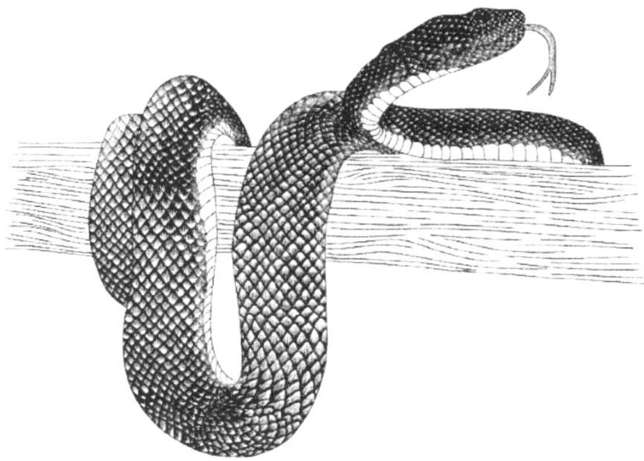

Figure 4.1: Brown tree snake [Illustration credit: Uzma Shaikh, TIFR Mumbai, India]

This invasion ultimately led to the extinction of twelve bird species native to the island's forests. Among these extinct species were the Guam rail (locally referred to as Kókó), Guam Kingfisher, Mariana fruit dove, Guam flycatcher, rufous fantail, Guam broadbill, and Micronesian myzomela. This incident underscores the vulnerability of isolated ecosystems to predators and serves as a stark reminder of the far-reaching impacts on wildlife.

## Lionfish: The new agent of death in the Atlantic

Shifting focus to the Atlantic, we encounter another invasive predator: the lionfish (*Pterois volitans* and *Pterois miles*). These fish are voracious hunters and possess venomous spines that incapacitate their prey. Originally from the Indo-Pacific region, lionfish have established themselves in the Caribbean and western Atlantic. Researchers Mark Albin (University of South Alabama) and Mark Hixon (the

University of Hawai'i at Mānoa) conducted studies suggesting that a singular lionfish residing within a coral reef could diminish native reef fish populations by up to 79 per cent within five weeks (Figure 4.2). But how did they end up so far from their Indo-Pacific home? There's a theory that in 1992, when Hurricane Andrew hit Florida, some lionfish escaped from aquariums. And now, they're living their best lives, wreaking havoc on local fish communities in the Atlantic Ocean.

Figure: Lionfish [Illustration credit: Uzma Shaikh, TIFR Mumbai, India]

From Guam's bird crisis to the lionfish invasion in the Atlantic, these stories remind us that nature can be unpredictable and invasive predators can turn a thriving ecosystem into a battlefield.

## Invasive species and human contribution:

### *Humans – the accidental villain!*

Humans have brought invasive predators into new environments through various means, both on purpose and by accident. As people travel and trade goods globally, they unintentionally transport invasive species. Take rats, for instance – black rats, Norway rats, and Polynesian rats have hitched rides on ships and popped up in many parts of the globe. Additionally, many scorpions and exotic spiders often travel far from home by sneaking into shipments of tropical fruits and vegetables. In a particular case, several policy makers proposed the introduction of cane toads from South America into Australia as a solution for beetle infestations; however, this led to the extinction of numerous native species. Additional invasive species introduced by humans include the Kudzu vine, Burmese python, house sparrows, starlings, and nutria. The aquarium industry and seafood markets can contribute to the movement of invasive species. Commercial ships carry ballast water and transport aquatic organisms like alien crabs and shrimps to new areas globally, while smaller boats might pick them up on their propellors. Zebra mussels are now an invasive species that made their way into the Great Lakes after ballast water from a foreign ship was dumped into Lake St. Clair in 1986. Another way invasive alien species can enter new areas is through canals. These man-made waterways link different bodies of water and create paths for ships and boats. The development of canal systems can facilitate fish and other species to migrate to new areas.

### *The Hawaiian tragedy:*

Humans often let their pets go, whether on purpose or by accident and sometimes, these critters become real troublemakers in the wild. There have been times when non-native species were introduced to help with pest issues, but they can become invasive threats that disrupt the local ecosystem. Take, for example, the case from 1865 when William Hillebrand thought it would be smart to bring myna birds (*Acridotheres tristis*) to Hawaii as a solution for armyworms and other insect nuisances. These birds are originally from India, the Himalayas, and the Middle East. Sadly, these mynas started competing with native bird species for food and nesting sites while also preying on the eggs and chicks of local birds. As a result, Hawaii has seen over 55 unique species disappear—mostly its native forest birds—with around 70% of those birds still alive now, classified as endangered.

## *Misadventure in New Zealand*

Another major tale of human-assisted biological invasion revolves around the stoat (*Mustela erminea*), a carnivore related to weasel and otter. Stoats go after a wide-ranging prey and climb trees to raid nests for eggs and chicks. Stoat will inspect each and every burrow to attack ground-nesting birds. The stoats were introduced from the Northern hemisphere to New Zealand in the 1880s, intended to control rabbit and hare populations – but it wreaked havoc on the local wildlife. Stoats are seen as the top threat to some of New Zealand's rarest birds and have played a role in wiping out several subspecies like the bush wren, laughing owl and New Zealand thrush. The stoats have also been linked to the kakapo disappearing from the mainland areas and likely contributing to the South Island kokapo's extinction. The sad tale of a massacre by stoats has been marked as one of the European colonists' biggest blunders in New Zealand's history.

## Love stories of parasites: Lethal affairs leading to extinction

Alien parasites often hitch a ride to new locations alongside an alien host. Once they arrive, these invaders switch from their original (non-native) host to a local one, starting a fresh host-parasite dynamic. Take the parasitic brood mite *Varroa destructor*, for instance. Its natural home is with the Asian bee species *Apis cerana*. Still, when European honeybees (*Apis mellifera*) showed up in Asia in the early twentieth century, they unknowingly picked up this mite along the way.

Another case involves tiny aquatic creatures called copepods. *Neoergasilus japonicus* is a parasitic copepod that originated in East Asia and has since spread to various parts of the globe, like Europe, America, and Africa. It can latch onto many types of freshwater fish as hosts. The main culprits behind its global spread seem to be aquaculture practices and the introduction of new fish species, which have pushed several local fish populations toward near extinction.

### *Mysterious disappearance of golden toad*

The golden toad (*Bufo periglenes*) became extinct and mysteriously disappeared before anyone could take stock of the crisis. In Panama, the golden toad is like a national treasure, a symbol of luck and prosperity. Many experts think these toads became extinct in their natural habitats largely because of deadly chytrid fungal infection and climate change-related issues. The chytrid fungus (*Batrachochytrium dendrobatidis*) causes a nasty disease called chytridiomycosis, which has been affecting amphibians all over the globe. Climate change seems to have given this fungus an edge in the golden toad's territory, making it even harder for them to survive. Climate change and associated seasonal shifts also interfered with golden toad's breeding habits. These toads relied on rainwater pools for mating and with adverse effects from climate change, which reduced rainfall and humidity levels – conditions crucial for breeding in rainwater pools. Climate change was favourable for the fungal parasite to infect golden toads. As a result, the entire toad population went extinct due to a lethal combo of climate change and fungal infection. By 2004, the golden toad was officially declared extinct, and its disappearance was more than just a sad tale; it marked the beginning of a bigger crisis for amphibians around the globe.

## The arrival of humans on Earth and its impact on web of life

### *Human-driven extinction*

Almost all species that have ever existed—about 99%—have gone extinct through five major extinction events, which were mostly triggered by natural causes like volcanic eruptions, shifts in tectonic plates, and asteroid impacts. But nowadays, the extinction we're seeing today is probably due to human actions, especially in the last hundred years or so. Experts say that anywhere between 100 and 10,000 species—ranging from tiny microbes to huge plants and animals—vanish yearly. That's a staggering extinction rate, about 100 to 1,000 times quicker than what we've seen historically. Species can disappear because people over-hunt and over-fish them, pollute their habitats, destroy where they live, or bring new species into places they don't belong. Often, when prey species face habitat destruction caused by humans, it can lead to their death or migration, affecting predator species and even driving them to extinction. Take the sabre-toothed tiger, for instance—this creature vanished around 8,000 to 10,000 years ago due to climate change and bone diseases. However, human encroachment on its territory played a significant role as well. However, human encroachment into its habitat coincided with the end of the ice age when its prey started diminishing, ultimately resulting in its final extinction (Figure 4.3).

Figure 4.3 : Saber-toothed tiger (extinct).

The hominins, the ancestors of modern humans, first appeared on Earth around 6 million years ago during the Miocene epoch. Humans started making a real impact on global biodiversity when modern *Homo sapiens* left Africa and spread to other continents. Hominins relied on hunting and gathering for food for much of their evolutionary journey. About 60,000 and 10,000 years ago, as people migrated into Eurasia, Australia, and the Americas, a wave of extinctions hit giant animals like mammoths, ground sloths, and kangaroos. Most of these megafaunas vanished mainly due to human hunting.

Figure 4.4: Woolly mammoth extinction occurred around 10,000 years ago.
[Illustration credit: Uzma Shaikh, TIFR Mumbai, India]

Figure 4.5: Woolly rhinoceros extinction occurred around 12,000 years ago.
[Illustration credit: Uzma Shaikh, TIFR Mumbai, India]

Woolly mammoths, closely related to elephants, faced extinction around 10,000 years ago from overhunting by humans and climate change—though some groups survived a bit longer (Figure 4.4). The woolly rhinoceros had a similar fate and disappeared between 12,000 and 14,000 years ago for the same reasons (Figure 4.5). The crested moa was one of nine moa species that once roamed New Zealand. All nine went extinct shortly after humans arrived in the mid-13th century because they couldn't handle even minimal hunting pressure (Figure 4.6).

Figure 4.6 : Crested moa, the flightless bird of New Zealand, became extinct between 1396 and 1442 AD. [*Illustration credit: Uzma Shaikh, TIFR Mumbai, India*]

The Irish elk was the biggest deer ever to exist, towering nearly 7 feet at the shoulder and tipping the scales at around 1,500 pounds. This massive creature roamed Europe, northern Asia, and North Africa (Figure 4.7). What stood out were its gigantic antlers, which could stretch up to a whopping 13 feet wide! The first major decline in their numbers happened about 12,000 years ago when they vanished from places like Ireland, Britain, and most of Europe.

Figure 4.7: Irish elk became extinct around 10,000 years.
[*Illustration credit: Uzma Shaikh, TIFR Mumbai, India*]

At the close of the Ice Age, things got super cold and harsh, leading to a major food shortage for the Irish elk deer. Their last population likely faced extinction due to climate changes warming their environment. As forests started taking over in this new climate, there was less open grassland for them to graze and move around with huge antlers. Humans might have also played a role in their decline. While we don't have much proof that people hunted them extensively, we do know that Neolithic settlements were spreading during this time—meaning hunting could have added more pressure on the already dwindling Irish elk population after environmental changes had made survival tough.

Figure 4.8: Dodo bird became extinct around 1690 on the island of Mauritius.
[*Illustration credit: Uzma Shaikh, TIFR Mumbai, India*]

Dodo bird (*Raphus cucullatus*) was native to Mauritius island in the Indian Ocean—about 500 miles off Madagascar's eastern coast. This forest-dwelling bird occasionally ventured near the beach. Dodos thrived there thanks to Mauritius's warm and rainy climate, with few natural predators and low resource competition (Figure 4.8). However, their adaptations made them unprepared for European sailors and the domesticated animals they brought along. Hunting them down, combined with invasive species like monkeys, deer, pigs, and rats, ultimately led to the dodo's extinction. After the arrival of settlers in Mauritius, the dodo vanished within a century. This extinction has profoundly affected the local ecosystem. Notably, it has diminished the capacity of Mauritian plants to disperse their seeds, potentially putting other species at risk of extinction as well. Overhunting, combined with the introduction of invasive species like monkeys, deer, pigs, and rats, ultimately led to the dodo's extinction

Great Auk was a flightless seabird killed by hunters for food and bait in the early 1800s. Hunters wiped out the great auk in huge quantities, frequently herding them up a plank and slaughtering them as they ascended into a ship's hold. These birds were completely defenceless and couldn't fly, which made it easy for hunters to capture them. The final recorded great auks were killed in June 1844 on Eldey Island, Iceland.

Figure 4.9: Passenger pigeons became extinct in the early twentieth century. The last survivor died at Cincinnati Zoo on September 1st, 1914.

[*Illustration credit: Uzma Shaikh, TIFR Mumbai, India*]

The passenger pigeon, once one of the most plentiful birds around, sadly vanished due to excessive hunting and loss of habitat (Figure 4.9). Native Americans hunted these pigeons, but things ramped up after Europeans showed up, especially during the 19th century. Pigeon meat became a cheap food source, leading to mass hunts for decades. There were even shooting tournaments where folks would take down hundreds of birds over just a few days! As cities expanded, the demand for pigeon meat skyrocketed; back in 1855 alone, about 300,000 pigeons were captured in the Midwest and shipped off to New York City. On top of all this hunting pressure came deforestation during European colonization—another nail in the coffin for these birds since they thrived on large stretches of uninterrupted forest land. The last wild passenger pigeon was shot near St. Vincent in Quebec in 1907. And then there's Martha—the very last known passenger pigeon—who lived out her final days at Cincinnati Zoo before passing away on September 1st, 1914; she had lost her two male companions four years earlier.

## IUCN: Monitoring pulse of life on Earth

An animal is labelled as endangered or critically endangered when its population becomes so small that it faces the risk of extinction—vanishing entirely. The International Union for the Conservation of Nature (IUCN) is the primary global authority for assessing species conservation status ( https://iucn.org/). Information regarding species that have become extinct since 1500 is compiled in the IUCN Red List, which offers comprehensive data for vertebrates, notably birds, mammals, and amphibians.

## The Red Data book

The Red Data Book, commonly referred to as the IUCN Red List, is a publicly accessible document that catalogues and disseminates information regarding the conservation status of endangered, rare, and vulnerable species across various categories, including plants, animals, and fungi (https://www.iucnredlist.org/search ).

The Red Data Book was established in 1964 by the IUCN, and it is an inventory of the global conservation status and extinction risk of biological species. IUCN conducts periodic evaluations of species to keep track of their status, and the IUCN Red List categorizes species into several classifications, including:

- *Extinct (EX):* A species is classified as extinct when there is no reasonable doubt that the last individual has perished.
- *Extinct in the Wild (EW):* This designation applies to species that only exist in captivity or in populations artificially maintained outside their natural habitat.
- *Critically Endangered (CR):* Species facing an extremely high risk of extinction fall into this category.
- *Endangered (EN):* This classification includes species with a significant risk of extinction.
- *Vulnerable (VU):* Species at a high risk of becoming endangered are categorized here.
- *Near Threatened (NT):* This group encompasses species on the brink of being classified as endangered in the foreseeable future.
- *Lower Risk (LR):* Species deemed unlikely to become endangered or extinct soon are placed in this category.

(a) (b)

Figure 4.10: (a) Pyranean Ibex, the mountain goat, officially declared extinct on Jan 6, 2000 [*Illustration credit: Uzma Shaikh, TIFR Mumbai, India*] (b) A subspecies of Pyranean Ibex, *Capra pyrenaica pyrenaica*, used to be an abundant animal in Spain.

## Traders & killers of wildlife

Over the last hundred years, wildlife smuggling and poaching have led to the extinction of about 20 species, like the Pyrenean Ibex (Figure 4.10), West African Black Rhinoceros (Figure 4.11), and Passenger Pigeon (Figure 4.9). These illegal activities are still widespread today, putting many creatures, such as lemurs, orangutans, pangolins, and others, at risk of disappearing forever.

Figure 4.11: The West African black rhinoceros (*Diceros bicornis longipes)* was declared extinct in 2011 by IUCN. [*Illustration credit: Uzma Shaikh, TIFR Mumbai, India*]

The rising interest in so-called "folk medicines" or "Traditional Chinese Medicinal" products has disastrously impacted various wildlife species. In many instances, hunting animals for their body parts to make traditional medicine is a major factor pushing them towards extinction. The high value placed on rare items like rhino horns, elephant ivory, and tiger parts fuels this illegal wildlife trade. For example, tiger bones and nearly all parts of their bodies are used in Traditional Chinese Medicine for various claimed benefits, including pain relief. While the Convention on International Trade in Endangered Species of Wild Fauna and Flora (CITES) aims to regulate international trade of certain species, its safeguards aren't sufficient to protect at-risk populations.

Some animals urgently need measures to shield them from human conflict, poachers, smugglers, and habitat destruction. For ongoing updates and to access a comprehensive list of recently extinct species and information on critically endangered, endangered, and threatened species, it is advisable to visit the IUCN website (at https://www.iucnredlist.org/search). Illustrative figures of only a few selected animals became extinct over the last half a century (Figures 4.12; 4.13; 4.14; 4.15). , alongside those currently classified as endangered or critically endangered (Figures 4.16; 4.17), are featured in this chapter.

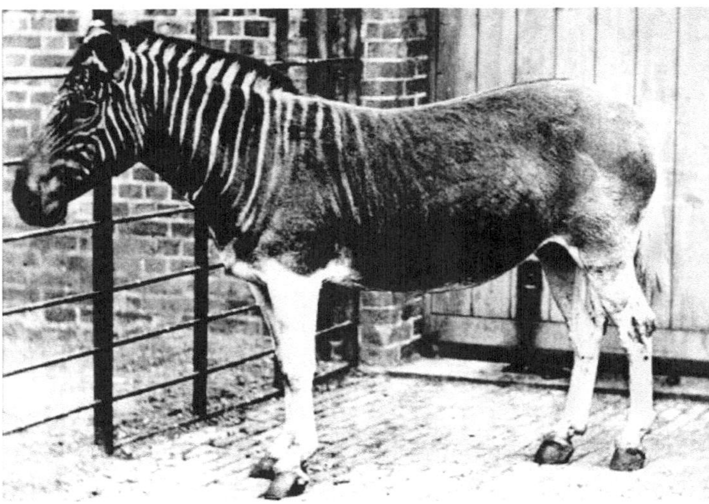

Figure 4.12: Quagga, photographed in London Zoo in 1870. The quagga was a subspecies of the plains zebra that inhabited South Africa until it was driven to extinction by hunting practices in the late 1800s. The last quagga died in the Amsterdam Zoo on Aug.12, 1883.

Figure 4.13: Tasmanian tiger, also known as the thylacine. Tasmanian tiger was found in Papua New Guinea, Tasmania, and throughout the Australian mainland. The last known animal was photographed in 1933 at Berlin Zoo.

Figure 4.14: Saudi gazelle (*Gazella saudiya*) became extinct in the wild: It was listed as extinct on the IUCN Red List in 2008. [*Illustration credit: Uzma Shaikh, TIFR Mumbai, India*].

Figure 4.15: The Japanese otter was officially declared extinct by the Ministry of Environment in Japan on August 28, 2012. [*Illustration credit: Uzma Shaikh, TIFR Mumbai, India*].

Figure 4.16: American Ivory Billed Woodpecker became extinct in the twenty-first century. The bird was announced extinct on 29th September 2021 by the US Fish and Wildlife Service. [*Illustration credit: Uzma Shaikh, TIFR Mumbai, India*].

Figure 4.17: Amur leopard — one of the most critically endangered animals, and in 2024, only around 100 Amur leopards have remained alive in the wild.
[*Illustration credit: Uzma Shaikh, TIFR Mumbai, India*].

Figure 4.18: Critically endangered lowland Gorilla (whereas the mountain Gorillas are classified as endangered by IUCN) [Photo credit: Dušan Veverkolog on Unsplash].

*Gorilla:* Gorillas are primarily herbivorous, ground-dwelling great apes residing in equatorial Africa's tropical forests (Figure 4.18). As the largest species among great apes, gorillas construct nests for sleeping—typically on the ground but occasionally in trees. Unfortunately, they face threats from hunting for their meat and capture of infants for the pet trade. Additionally, poaching poses a risk by facilitating the transmission of diseases such as Ebola to humans. Over the past years, these factors have contributed to a decline in gorilla populations exceeding 60%. The International Union for Conservation of Nature (IUCN) has classified gorillas as endangered on its Red List. Efforts by organizations like the World Wildlife Fund (WWF) focus on ensuring their protection.

Figure 4.19: Critically endangered orangutans, [Photo credit: Dimitry B. on Unsplash].

*Orangutans:* It's crucial to focus on all three orangutan species—the Sumatran orangutan, Bornean orangutan, and Tapanuli orangutan—each critically endangered. These great apes are only found in Indonesia and Malaysia's rainforests; their range once extended across Southeast Asia and South China during the Pleistocene era. People often hunt orangutans for bushmeat or traditional medicine use. They also face threats from habitat loss due to logging and palm oil farming (Figure 4.19).

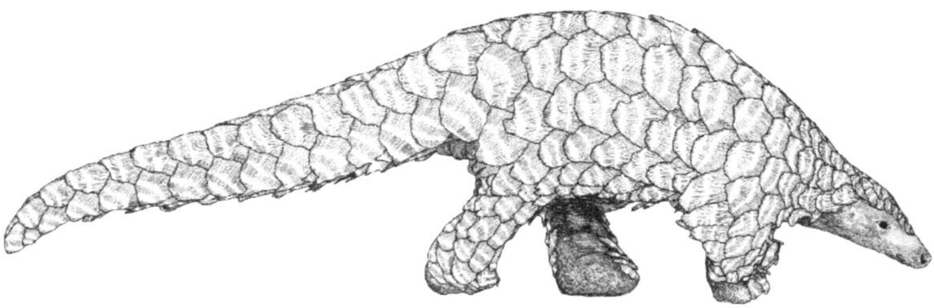

Figure 4.20: Pangolin: All species of pangolins face the threat of extinction, with several (Sunda, Philippine and Chinese pangolins) categorized as critically endangered.
[*Illustration credit: Uzma Shaikh, TIFR Mumbai, India*].

*Pangolins:* Pangolins are, also known as scaly anteaters, are mammals facing severe risks from poaching and habitat loss due to demand for their meat and scales (Figure 4.20). They hold the title of the most trafficked wild mammal globally. Their meat is considered a delicacy in Asia, while their scales fetch high prices on the black market for use in traditional medicine. There's even a market in the U.S. for pangolin leather used in boots, bags, and belts! Despite no scientific evidence supporting any medicinal claims—like curing arthritis or enhancing male vitality—people still pursue these products aggressively.

*Lemurs:* Lemurs are the oldest living primates on Earth (dating back over 70 million years!), and they originated long before humans existed when lemur-like creatures roamed Africa alongside dinosaurs (Figure 4.21). Native only to Madagascar (with some types found on nearby islands like Comoros), lemurs are recognized by the International Union for Conservation of Nature (IUCN) as one of the most endangered mammals worldwide. Alarmingly, 98% of all lemur species face threats that could lead to extinction, with a staggering 31% classified as critically endangered. They often get hunted for food or captured due to cultural beliefs or demands in the pet trade—even though it's illegal! The main threats include habitat destruction from slash-and-burn agriculture, illegal logging, and wildfires.

Figure 4.19: Lemurs: The most endangered mammals, [Photo credit: Zdeněk Macháček on Unsplash].

## Halting the extinction crisis

The risk of extinction has become increasingly prevalent, and many species are hanging by a thread. Existing threats and the effectiveness of protective measures influence the impending extinction rate. Endangered species may ultimately face extinction if they do not receive adequate protection. Animals and plants are vanishing from our planet at a shocking pace. A United Nations report warns that around one million species face the risk of extinction, with projections suggesting that over half of all species could disappear by 2050. The preservation of endangered species is not merely an ethical obligation; it is essential for maintaining the health of our environment. Even though we're facing an uncertain future due to climate change and habitat destruction, there's still hope for these vulnerable species. Thanks to various organisations' initiatives and conservation measures, numerous species are experiencing or have already achieved recovery. The committed work of entities such as the International Union for Conservation of Nature (IUCN), World Wildlife Fund (WWF), Center for Biological Diversity, International Anti-Poaching Foundation, Wildlife Alliance, The World Society for the Protection of Animals, International Fund for Animal Welfare, Oceana, Jane Goodall Institute, Environmental Investigation Agency, and Wildlife Conservation Society is crucial in protecting endangered species. Their multifaceted approaches encompass habitat preservation, advocacy for policy changes, community involvement efforts, and direct interventions against poaching and illegal wildlife trade.

"The Crocodile Hunter", Stephen Robert Irwin, the famous Australian conservationist, wildlife educator, and environmentalist, once said "We don't own the planet Earth; we belong to it. And we must share it with our wildlife."

# CHAPTER 5

# *Future of Life*

For centuries, humanity has pondered the ultimate fate of all life on our planet. Countless scientific and spiritual theories offer insights into how this journey may conclude. However, the stark reality is that our planet is already in trouble and has been suffering for an extended period. Earth's past gives us insight into changes no one has witnessed and helps us connect them to what's happening today. This final chapter of the book highlights how Earth has sustained life and alerts us to how future life may alter our planet. One key takeaway from the last few thousand years is the realization that human impact on Earth's systems has accelerated over time.

## How long will Earth exist?

No one knows how things will unfold or when life on Earth might end. If you take a trip through Earth's history and check out the first two chapters of this book, you will get a glimpse of how our planet came to be and how life started. However, our planet is destined to end someday. When our Sun dies, life on Earth will also end. The Sun is the main energy source, keeping everything alive here on Earth. Without it, we wouldn't even exist. At some point in the future, perhaps billions of years from now, the Sun will fizzle out as a white dwarf – a dead star that has exhausted all its hydrogen --- the nuclear fuel. Without the Sun, all plants would perish first (no photosynthesis and no food for themselves), followed by animals that rely on plants for food – including humans would die too. In short, all the food chains (linking organisms --- who eats whom!) and food webs (network of all food chains) will collapse.

Like a science fiction storyteller, I can fantasize that future descendants of humans might find ways to hang on in an AI-supported, high-tech, mechanized, insulated environment equipped with stored nutritional resources for days or months without sunlight. But no matter what our future descendants may do, sustaining life under these conditions will not be feasible forever. Now, someone may ask how much time Earth has before it is engulfed by the dying Sun. Experts have estimated we have a few billion years left. However, most life forms are likely to end much earlier than predicted due to rapid environmental changes shaking up ecosystems everywhere.

## Anthropocene "The age of human":

It's becoming increasingly clear that humans have had such a huge impact on our planet and all its different life forms that the changes we've made might last forever—and could even be impossible to reverse—affecting everything from the environment to biodiversity. Despite Earth's vast history spanning 4.5 billion years, modern humans have existed for only about 200,000 years. In this brief span, we have dramatically transformed the world's physical, chemical, and biological fabric upon which all living beings rely. We have become catalysts for significant transformations—accelerating global warming and altering landscapes, ecosystems, water sources, and the atmosphere. A growing cohort of scientists contends that the Anthropocene should succeed the Holocene Epoch (~11,700 years ago to present), suggesting a commencement year around the mid-20th century. "Anthropocene," derived from Greek, translates to "recent age of man."

## Are humans vexing nature and 'playing God'?

For the past 11,500 years, Earth has been in the Holocene epoch. This period kicked off after the last ice age, when glaciers melted away. During these years, we've seen major transformations on our planet - rapid population growth and modern societies popping up everywhere. Humans have constructed cities over this time and made massive strides in technology. Since those ice age days, Earth's climate has generally been warm and stable; however, that's changing fast, with rising temperatures disrupting that stability. Particularly, over the last six decades — often called the '*Great Acceleration*', human-induced changes have become conspicuous with unprecedented speed and magnitude. Scientists now agree that human actions are causing accelerated global warming, agriculture expansion, urban development, deforestation, and pollution, reshaping our planet like never before. Historically, there have been mass extinction events where many species vanished due to natural disasters like asteroid strikes or volcanic eruptions. But now we're witnessing something different: for the first time ever, one single species (*Homo sapiens* – the human) is wreaking havoc on nature with full awareness of what it's doing! Escalating carbon dioxide emissions, rising global temperatures, ocean acidification, habitat destruction, extinction events, and extensive natural resource exploitation vividly illustrate how deeply we have altered our planet's state.

Looking at where we're headed in the 21st century, it's pretty clear that what we do as humans greatly impacts Earth's environment and ecosystem. With rapid industrial growth and a growing population, the natural balance that has kept life thriving for ages is under serious stress. We're dealing with major issues like the degradation of ecosystems, climate change, loss of habitats, and depletion of resources. According to a report from Global Footprint Network in 2022, our ecological footprint now goes way beyond what the Earth can handle – basically, it can't keep up with how fast we're using natural resources (https://www.footprintnetwork.org/). Last year's global footprint was about 1.7 times greater than what the planet can regenerate in a year. So, Earth takes around 1.7 years to make back what we use in just one year!

## Global warming and climate change: Hype or reality?

From 2011 to 2020, we experienced the hottest decade, with global temperatures hitting 1.1°C above pre-industrial levels in 2019. Fast forward to September-November of 2024, and we're seeing a global surface temperature that's about 1.31°C (2.36°F) higher than the average of 14.0°C (57.1°F) recorded throughout the 20th century [https://www.ncei.noaa.gov/access/monitoring/monthly-report/global/202411]. The planet is warming at an alarming pace—around 0.2°C every ten years, reminding all the countries worldwide to realize that it's crucial to keep this rise well under 2°C, ideally aiming for just 1.5°C.

At the heart of climate change lies the greenhouse effect—a natural phenomenon in which certain gases like carbon dioxide, methane, nitrous oxide, and fluorinated gases act like glass panes in a greenhouse, trapping heat from sunlight and preventing it from escaping back into space—and hence the name 'greenhouse gas'! Human activities have ramped up these gas levels significantly; by 2020, $CO_2$ concentrations were already at about *48%* over those early industrial days (during the 1750s). Every ton of $CO_2$ released fuels global warming's fire, contributing further to environmental instability. Can we slow down this trend? Yes, we need worldwide net-zero $CO_2$ emissions across the board! Moreover, minimizing emissions of other potent greenhouse gases like methane can yield immediate benefits against rapid climatic shifts.

A heating planet bears significant repercussions across various facets within our shared environment—for instance—as temperatures climb ever higher—it leads glaciers not only to melt faster but also accumulate less snow over time, resulting ultimately in increased sea levels worldwide. The predictions anticipate mountain glacier melting could escalate overall sea level height between 0.28 to 1 meter (11–39 inches) by the year 2100! Rapid climate shift will lead to more frequent, prolonged, and severe hurricanes and cyclones characterized by intensified winds, storm surges, cloud bursts and elevated and unseasonal rainfall rates. Additionally, the ocean itself absorbs considerable amounts of $CO_2$ released, which induces seawater acidification, posing dire threats to marine ecosystems everywhere!

## Sixth mass extinction or an end of another beginning?

As our populations have surged and technology has advanced at an unprecedented rate, our impact on the natural world has intensified. Since modern humans first moved out of Africa, colonization and various human activities have sped up many species' loss, resulting in a biodiversity crisis. We have invaded—and sometimes completely eradicated—the habitats vital for many species' survival. Additionally, we have polluted our air, soil, freshwater sources, and oceans, altering atmospheric conditions and climate patterns profoundly. These rapid changes are disrupting Earth's systems on a global scale. Across continents, countless species are dwindling due to many harmful human activities. These include habitat destruction and fragmentation, extensive use of pesticides and other harmful chemicals, overhunting and exploitation, and introduction of invasive species. Research indicates that current extinction rates are alarmingly between 1,000 and 10,000 times higher than the natural background levels, prompting experts to assert that our planet is indeed experiencing its sixth mass extinction this time. However, humanity itself stands at its core.

Life on Earth has consistently demonstrated a remarkable capacity for recovery, even after experiencing five consecutive mass extinction events. Extinction and evolution go hand in hand; you can't have one without the other. Should another mass extinction occur, there would be a considerable gap—spanning millions of years in geological terms—before life begins to flourish again. If the current extinction rate persists and drastically impacts Earth's biodiversity, the hardy survivors (which may include bacteria or various microbes) will likely form the foundational population for future ecosystems. Thus, the cycle of evolution will restart anew, suggesting that "life never truly ceases on Earth."

## What does the future look like? Can we prevent climate apocalypse?

While we can't stop climate change completely, we can surely slow it down. Like global warming, which won't disappear in a day, we can still work on reducing its pace and keeping it in check. The key is cutting back on human emissions of carbon dioxide and other greenhouse gases that really drive temperatures up around the planet.

As we burn fossil fuels and cut down forests, high concentrations of greenhouse gases, specifically carbon dioxide, threaten to raise the planet's average surface temperature to intolerable levels. We already see effects scientists predicted, such as the loss of sea ice, melting glaciers and ice sheets, sea level rise, and more intense heat waves. Scientists predict that the rise in global temperature due to greenhouse gases will continue and severe weather damage will also intensify.

By 2050, 800 million people will live in cities where sea levels could rise by more than half a metre. Many coastal communities worldwide are already facing the threat of rising sea levels and coastal flooding, and climate impacts can drown their neighbourhoods soon.

Figure 5.1 Green sea turtle: (Photo credit: Zdeněk Macháček on Unsplash)

## Recovering biodiversity --- reviving life for the future

Human activities often dominate discussions for their destructive effects on the environment; it is important to acknowledge whatever positive contributions are also being made. The sustainable practices advocated by the United

Nations through its 17 Sustainable Development Goals (https://sdgs.un.org/goals) aim to protect our planet and its life forms. Initiatives such as *in situ* conservation (*species conserved within their natural environment*) efforts alongside *ex situ* strategies (*conservation of species outside their native environment under a human-controlled environment*) implemented by various organizations, combined with technological innovations, promise to mitigate environmental degradation and even rejuvenate ecosystems. For instance, initiatives like reforestation projects, adopting renewable energy sources, and advancing eco-friendly methodologies are making considerable progress towards lessening our adverse effects on nature. These green efforts have aimed to compensate for biodiversity loss and contribute to carbon sequestration, helping mitigate climate change.

Figure 5.2 Giant panda. (Photo Credit: Zdeněk Macháček on Unsplash)

It is worth sharing many success stories portrayed by the National Wildlife Federation (https://www.nwf.org) in their narratives; "*when we invest in wildlife conservation, wildlife wins*". Ongoing concerted efforts in many countries, governmental and nongovernmental organizations have brought back numerous species from the brink of extinction, including the Bald eagle, California condor, Swift fox, Foskett speckled dace, Karner blue butterfly, Kirtland's warbler, Licking river mussels, Louisiana black bear, Monitor Gecko, New England cottontail, Delmarva fox squirrel, Trumpeter swan, American alligator, Green sea turtle (Figure 5.1), Piping plover, Peregrine falcon, Channel island fox, Humpback whale, Puerto Rico parrot, Whooping crane, Steller sea lion, Greater one-horned rhino, Azores butterfly, Giant panda (Figure 5.2) and many more listed under the "Green status of species"

(https://iucn.org/resources/conservation-tool/iucn-red-list-threatened-species#5757).

Figure 5.3 Polar bear (Photo credit: Brian McMahon on Unsplash)

Nevertheless, we cannot rejoice in conservation success stories without turning to polar bears, which are endemic to arctic regions and majestic apex predators adept at surviving extreme cold temperatures (Figure 5.3). Sadly, they now face a drastic decline due to climate change, pollution, and energy development activities that disrupt their habitats. Through these conservation success stories—alongside the urgent calls for action of many critically endangered animals like polar bears—we are reminded of our collective effort to protect our planet's precious wildlife.

## Epilogue

Things are only going to get tougher if climate change keeps ramping up. Looking back at what has happened on our planet and how life has adapted to changes, we can make smarter predictions and plans for our future. Our collections and scientific studies aim to show where we stand today and what actions we can take for a brighter tomorrow. We all play a role in safeguarding our planet and its wildlife for future generations.

When non-avian dinosaurs went extinct around 66 million years ago, mammals took advantage and filled many ecological gaps left behind. If humans were to vanish today, it's hard to say if another species would evolve to reach human-level intelligence or take over as the most capable land animals. However, predicting how evolution will play out over the next few million years remains mostly hypothetical!

# *Bibliography*

Atkins J.F., Gesteland R.F. & Cech T.R. *RNA Worlds: From Life's origins to diversity in gene regulation.* Cold Spring Harbor Laboratory Press, 2010.

Attenborough D. *A life on our planet: My witness statement & a vision for the future.* Grand Central Publishing, 2020.

Benton M.J. *Extinctions: How life survives, adapts and evolves.* Thomas & Hudson, 2023.

Benton M.J. *When life nearly died: The greatest mass extinction of all time.* Thomas & Hudson, 2002.

Black R. *The last days of the Dinosaurs: An asteroid, extinction and the beginning of our world.* St. Martin's Press, 2022.

Brusatte S. *The rise and fall of the Dinosaurs: A new history of a lost world.* Mariner Book, 2018.

Cairns-Smith A.G. *Seven clues to the origin of life: A scientific detective story.* Cambridge University Press, 1990.

Chinery M. *Animal Kingdom: Life in the wild.* Anness, 2011.

Dawkins R. The ancestor's tale: A pilgrimage to the dawn of evolution. Mariner Books, 2005.

Dawson A. *Extinction: A radical history.* QR Books, 2016.

Deamer D.W. & Szostak J.W. *The Origins of Life.* Cold Spring Harbor Laboratory Press, 2010.

Deamer D.W. & Thorne S.R. Assembling life: How can life begin on Earth and other habitable planets? Highbridge Audio & Blackstone Publ. 2021.

Deamer D.W. Origin of life: What everyone needs to know. Oxford University Press, 2020.

Dyson F. Origins of life. (2nd edition). Cambridge University Press. 1999.

Edwards L.E. & Pojeta Jr. Fossils, rocks and time. US Geological Survey, 1999.

Eldridge N. Extinction and evolution: What fossils reveal about the history of life. Firefly Books, 2014.

Felbab-Brown V. The extinction market, wildlife trafficking and how to counter it. C. Hurst & Co. Publishers Ltd., 2017.

Fortey R. Fossils: The key to the past. Smithsonian Institution Press, 2002.

Frankel C. Extinctions: From Dinosaurs to you. University of Chicago Press, 2024.

Gates B. How to avoid a climate disaster: The solutions we have and the breakthrough we need. Knopf Publisher, 2023.

Goodall J., Hudson G. & Thane M. Hope for animals and their world: How endangered species are being rescued from the brink. 2009, Grand Central, 2009.

Halliday T. Otherlands: A journey through Earth's extinct worlds. Random House, 2022.

Joyce G.F. The RNA World: Life before DNA and Protein. Cambridge University Press, 2009.

Knapp, E.J. Dead serious: Wild hope amid the sixth extinction. Torrey House Press, 2021.

Knoll A. Life on a young planet: The first three billion years of evolution on Earth. Princeton University Press, 2015.

Kolbert E. The sixth extinction: An unnatural history. Henry Holt & Co., 2014.

Livio M. & Szostak J. Is Earth exceptional? The quest for cosmic life. Basic Books, 2024.

Mackay R. The atlas of endangered species. Routledge, 2017.

Margulis L. & Chapman M.J. Kingdoms and domains: An illustrated guide to the phyla of life on Earth. (4th edition). Elsevier Inc., 2010.

Neubeck A. & McMahon S. Prebiotic chemistry and the origin of life. Springer, 2021.

Oparin A.I. The chemical origin of life. Charles C. Thomas, 1964.

Pobst S. & Fuller T.K. Animals on the edge: Science races to save species threatened with extinction. National Geographic, 2008.

Propen A.D. At home in the Anthropocene. The Ohio State University Press, 2022.

Rauchfuss H. & Michell T.N. *Chemical evolution and the origin of life*. Springer, 2008.

Redfern R. *The evolution of the continents, Oceans and life*. University of Oklahoma Press, 2001.

Rich N. *Losing Earth: A recent history*. MCD Publications, 2019.

Ritchie H. *Not the end of the world: How we can be the first generation to build a sustainable planet.* Little Brown Spark, 2024.

Schaller G.B. *The Last Panda.* University of Chicago Press, 1993.

Schlossman M. Extinction: *Our fragile relation with life on Earth.* Ammonite Press, 2022.

Schopf J. W. *Life's origin: The beginnings of biological evolution.* University of California Press, 2002.

Schopf J.W. *Cradle of life: The discovery of Earth's earliest fossils.* Princeton University, 2021.

Schreiber U.C. & Mayer C. *The first cell: The mystery surrounding the beginning of life.* Springer, 2020.

Secher A., *Travels with Trilobites.* Columbia University Press, 2022.

Spears P. *Kingdoms of life connected.* (3rd edition) Conifer, Colo.: Big Picture Science, 2022.

Thunberg G. *The climate book: The facts and the solutions.* Penguin Press, 2023.

Tudge C. *Last animals at the Zoo: How mass extinction can be stopped.* Island Press, 1992.

Vié J.C., Hilton-Taylor C. & Stuart S.N. (editors) *Wildlife in a changing world: An analysis of the 2008 IUCN Red List of Threatened species,* IUCN 2008.

Vince G. *Adventures in the Anthropocene: A journey to the heart of the planet we made.*

Milkweed Editions, 2014.

Walker C. & Ward D. *Fossils.* (revised edition) Smithsonian Handbooks, 2002.

Wallace-Wells D. *The uninhabitable Earth: A story of the future.* Tim Duggan Books, 2019.

Wilson E. O. *The loss of biodiversity is a tragedy.* Entrevista à UNESCO - United Nations Educational, Scientific and Cultural Organization, 2010.

Winter M. (2019) *Wildlife conservation: Principles and practices.* Callisto Reference, 2019.